ESSENTIAL SKILLS
FOR GCSE
Physics

Roy White

HODDER
EDUCATION
AN HACHETTE UK COMPANY

Although every effort has been made to ensure that website addresses are correct at time of going to press, Hodder Education cannot be held responsible for the content of any website mentioned in this book. It is sometimes possible to find a relocated web page by typing in the address of the home page for a website in the URL window of your browser.

Hachette UK's policy is to use papers that are natural, renewable and recyclable products and made from wood grown in well-managed forests and other controlled sources. The logging and manufacturing processes are expected to conform to the environmental regulations of the country of origin.

Orders: please contact Bookpoint Ltd, 130 Park Drive, Milton Park, Abingdon, Oxon OX14 4SE. Telephone: +44 (0)1235 827827. Fax: +44 (0)1235 400401. Email education@bookpoint.co.uk Lines are open from 9 a.m. to 5 p.m., Monday to Saturday, with a 24-hour message answering service. You can also order through our website: www.hoddereducation.co.uk

ISBN: 978 1 510 46002 7

© Roy White 2019

First published in 2019 by
Hodder Education,
An Hachette UK Company
Carmelite House
50 Victoria Embankment
London EC4Y 0DZ

www.hoddereducation.co.uk

Impression number 10 9 8 7 6 5 4 3 2 1

Year   2023 2022 2021 2020 2019

Cover photo © kotoffei - stock.adobe.com

Typeset by Integra Software Services Pvt. Ltd., Pondicherry, India

Printed by Replika Press Pvt. Ltd., Haryana, India

A catalogue record for this title is available from the British Library.

Contents

❯❯ How to use this book

Welcome to *Essential Skills for GCSE Physics*. This book covers the major UK exam boards for Science: AQA, Edexcel (including Edexcel International GCSE), OCR 21st Century and Gateway, WJEC /Eduqas and CCEA. Where exam board requirements differ, these specifics are flagged. This book is designed to help you go beyond the subject-specific knowledge and develop the underlying essential skills needed to do well in GCSE Science. These skills include Maths, Literacy, and Working Scientifically, which now have an increased focus.

- The Maths chapter covers the five key areas required by the government, with different Physics-specific contexts. In your Physics exams, questions testing Maths Skills make up 30% of the marks available.
- The Literacy chapter will help you learn how to answer extended response questions. You will be expected to answer at least one of these per paper, depending on your specification and they are usually worth six marks.
- The Working Scientifically chapter covers the four key areas that are required in all GCSE sciences.
- The Revision chapter explains how to improve the efficiency of your revision using retrieval practice techniques.
- The Exam Skills chapter explains way of improving your performance in the actual exam.

To help you practise your skills, there is an exam-style paper at the end of the book, with another available online at www.hoddereducation.co.uk/EssentialSkillsPhysics. While they are not designed to be accurate representations of any particular specification or exam paper, they are made up of exam-style questions and will require you to put your maths, literacy and practical skills into action.

Key features

In addition to Key term and **Tip** boxes throughout the book, there are several other features designed to help you develop your skills.

Ⓐ Worked examples

These boxes contain questions where the working required to reach the correct answer has been shown.

Ⓐ Expert commentary

These sample extended responses are provided with expert commentary, a mark and an explanation of why it was awarded.

Ⓑ Guided questions

These boxes guide you in the right direction, so you can work towards solving the question yourself.

Ⓑ Peer assessment

These activities ask you to use a mark scheme to assess the sample answer and justify your score.

Ⓒ Practice questions

These exam-style questions will test your understanding of the subject.

Ⓒ Improve the answer

These activities ask you to rewrite the sample answer to improve it and earn full marks.

Answers to all questions can be found at the back of the book. These are fully worked solutions with step-by-step calculations included. Answers for the second online exam-style paper can also be found online at www.hoddereducation.co.uk/EssentialSkillsPhysics.

★ **Flags like this one will inform you of any specific exam board requirements.**

1 Maths

To do well in GCSE Physics you will need to be familiar with the units in which physics measurements are made and how they are combined mathematically. Perhaps more than in any other GCSE science, the development of your mathematical skills in physics is vital if you are to achieve the highest grades.

>> Units and abbreviations

Physicists use the SI system of measurement. This system is based on the seven fundamental base units shown in Table 1.1.

Table 1.1 **Base units in GCSE Physics**

Measurement	Unit	Abbreviation
mass	kilogram	kg
length	metre	m
time	second	s
current	ampere	A
temperature	degree Celsius	°C
amount of substance	mole	mol
candela	luminous intensity	cd

All other SI units are combinations of the base units. These combinations are called derived units (see Table 1.3).

It would be inappropriate to give, say, the mass of a postcard in kilograms, so physicists use smaller (submultiple) and larger (multiple) units in calculations. The common ones are shown in Table 1.2.

Table 1.2 **Submultiple and multiple units in GCSE Physics**

Prefix	Symbol	Meaning	Example
micro	µ	one millionth	1 µA = 1/1 000 000th of an Amp
milli	m	one thousandth	1 ms = 1/1000th of a second
centi	c	one hundredth	1 cm = 1/100th of a metre
kilo	k	one thousand	1 kg = 1000 grams
Mega	M	one million	1 Mm = 1 000 000 m or 1000 km

These examples show sensible units to measure the objects provided:

- A postcard measures 15 cm by 8 cm.
- The distance between London and Birmingham is about 190 km.
- The diameter of a one pound coin is 22.5 mm.

> **Tip**
>
> There are also a few physical quantities that are simple ratios and have no unit. At GCSE these are:
> - efficiency
> - magnification
> - transformer turns ratio.
> Remember you do **not** need to put a unit symbol with these quantities.

> **Key terms**
>
> Base units: The units on which the SI system is based.
>
> Derived units: Combinations of base units such as m/s and kg/m^3.
>
> Submultiples: Fractions of a base unit or derived unit, such as centi- in centimetre.
>
> Multiples: Large numbers of base or derived units, such as kilo- in kilogram.

> **Tip**
>
> SI stands for Système International. You don't need to know about the history of the system – you just need to know the units.

Table 1.3 GCSE Derived units

Physical quantity	Derived unit	Abbreviation
area	square metres	m^2
volume	cubic metres	m^3
density	kilogram per cubic metre	kg/m^3
temperature	degree Celsius	°C
pressure	pascal	Pa
specific heat capacity	joule per kilogram per degree Celsius	J/kg/°C
specific latent heat	joule per kilogram	J/kg
speed	metre per second	m/s
force	newton	N
gravitational field strength	newton per kilogram	N/kg
acceleration	metre per squared second	m/s^2
frequency	hertz	Hz
energy	joule	J
power	watt	W
electric charge	coulomb	C
electric potential difference	volt	V
electric resistance	ohm	Ω
magnetic flux density	tesla	T

Tip

Above 9999, groups of digits are usually separated in groups of three by a space; for example, 10 000. You should try to avoid using a comma to separate groups of digits because, outside the UK and the USA, a comma is often used as a decimal point. Mistaking a comma as the position of a decimal point when administering drugs or designing a bridge could be disastrous.

➤➤ Arithmetic and numerical computation

Expressions in decimal form

Decimal places look like full stops (.) in the middle of numbers. They are used to signify a value between two 'whole' numbers or integers. Any number can be written with a decimal place, even a whole number. For example, 2 can also be written 2.0 (one decimal place), 2.00 (two decimal places) and so on. A lot of calculations in physics result in answers that don't give a whole number and therefore you may have to use decimals.

For example, suppose we carry out a calculation and get an answer of 0.6547 m/s. This is an answer that has four decimal places (in other words, four numbers beyond the decimal point). But the question may specify that we give the answer to three decimal places (3 dp). In order to do that, we look at the decider figure and round up if it is more than four.

- So, if you need to give the number 0.6547 to 3 dp, the decider figure is the *fourth* figure after the decimal point (one after the number of decimal places needed). In 0.6547, the decider figure is 7. Since 7 is more than 4, we round up by adding a 1 to the third decimal place. So, to 3 dp, we write 0.6547 as 0.655.
- If we take the same number and work it out to 2 dp, the decider figure is the third figure after the decimal point (4). So, we do not round up and we write 0.6547 as 0.65.
- If we take the same number and work it out to 1 dp, the decider figure is the second figure after the decimal point (5). So, we round up the 6 to a 7 and 0.6547 is written as 0.7.

Key terms

Decimal places: The number of integers given after a decimal point.

Integers: These are whole numbers, which includes zeros.

Decider figure: The integer after the number of decimal places required which decides whether we must round up or not.

A Worked examples

1 In a circuit, the voltage across a device is 12.4 V and the current through it is 1.7 A. You work out that the resistance is 7.2941 Ω. Write this answer to 1 dp, 2 dp and 3 dp.

To 1 dp The decider is the second figure after the decimal point, which is a 9.

So, we round up to 7.3 (1 dp).

To 2 dp The decider is the third figure after the decimal point, which is a 4.

So, we do not round up.

We write 7.2941 as 7.29 (2 dp).

To 3 dp The decider is the fourth figure after the decimal point, which is a 1.

So, we do not round up.

We write 7.2941 as 7.294 (3 dp).

2 A physicist measures the diameter of a metal rod as 0.7 cm to 1 dp.

Giving both answers to 2 dp, what is:

a **the smallest diameter that the rod could have had?**

Step 1 The smallest number would still have to be a value that the physicist rounded up, so the first decimal place would be a 6.

Step 2 The number following the 6 would have to be as small as possible, but still a number that allows us to round up.

Step 3 So, the smallest diameter is 0.65 cm.

b **the largest diameter that the rod could have had?**

Step 1 The largest number must be small enough to not have been rounded up, so the first decimal place would be a 7.

Step 2 The number following the 7 would have to be as large as possible, but still a number that won't allow us to round up.

Step 3 So, the largest diameter is 0.74 cm.

> **Tip**
>
> Some answers may be recurring. In other words, the last number repeats forever. An example of this is if you divide the number 2 by the number 3 on your calculator. With this calculation, you may see on your display 0.6666666666, although more modern calculators display numbers like 0.6666666666... as 0.6̇. The dot above the 6 shows that the 6 repeats for ever. In either case, the rule for rounding is the same. For 3 dp, we would write 0.6666666666 or 0.6̇ as 0.667, and so on. Note that, if there are two dots, then the numbers *between* the dots are repeated. So, 0.6̇52̇ means 0.652652652... This would be 0.7 to 1 dp, 0.65 to 2 dp and 0.653 to 3 dp.

B Guided questions

1 A steel block is in the form of a cuboid. A physicist finds its dimensions are 1.21 cm × 3.42 cm × 5.63 cm. Calculate its volume, in cm³, to 2 dp.

Step 1 Find the volume: v = 1.21 cm × 3.42 cm × 5.63 cm = ... cm³

Step 2 Give the volume to 2 dp: volume = ... cm³

2 A toy car rolls down a slope and travels 20 cm in 6.4 seconds. Calculate the speed of the car in cm/s to 2 dp.

Step 1 Find the speed: $\frac{20\,cm}{6.4\,s}$ = ... cm/s

Step 2 Give the answer to 2 dp: speed of car = ... cm/s

> **Key term**
>
> Recurring: When a number goes on forever.

C Practice questions

3 An A4 sheet of paper measures 210 mm wide and 297 mm long. Each measurement is given to the nearest mm. Write down the maximum and minimum dimensions of the A4 sheet of paper, giving your answers in mm to 1 dp.

4 A steel rod has a rectangular cross section of 12.2 mm × 15.3 mm. Find its area of cross section giving your answer in mm^2 to 1 dp.

5 A student weighs 630 N and the total area of his feet in contact with the ground is 205 cm^2. Calculate the pressure he exerts on the ground, giving your answer in N/cm^2 to 1 dp.

Tip

Remember that to find the area of a rectangle, you multiply length by width.
To work out pressure, the formula is $p = \dfrac{F}{A}$.

Expressions in standard form

Physicists sometimes deal with very large or very small numbers. For example, the number of water molecules in a tablespoon of water is about 602 000 000 000 000 000 000 000 (an incredibly large number).

On the other hand, the wavelength of an X-ray is about 0.000 000 000 1 metres (an incredibly small number).

So, we need a better way to write down very large and very small numbers. We do that using powers of 10.

Powers of 10

Powers of 10 give us a way to write these very large and very small numbers in a sort of short-hand format. For example, if we look at the calculation $10 \times 10 = 100$, we can see that two tens are multiplied together. We can, therefore, write the value of 100 as 10^2 or 1.0×10^2.

In the calculation $10 \times 10 \times 10 \times 10 = 10000$, we can see that four tens are multiplied together. We can write the value of 10 000 as 10^4, or 1.0×10^4. The reason for the 1.0 in the second version will become clearer when standard form is covered in more detail on the following page.

Key term

Standard form: A number in the form a × 10^n used when writing down very large or very small numbers.

Bigger numbers than these examples are written in a similar way. This is summarised in Table 1.4.

Table 1.4 Positive powers of 10

Number	Written	Often written
10	1×10^1	10
100	1×10^2	10^2
1000	1×10^3	10^3
10 000	1×10^4	10^4
100 000	1×10^5	10^5
1 000 000	1×10^6	10^6

Numbers less than 1 are all written with negative indices as summarised in Table 1.5.

Key term

Index: Index is the power to which a number or letter is raised. The plural of index is indices.

Table 1.5 Negative powers of 10

Fraction	Decimal	Written	Often written
$\frac{1}{10}$	0.1	1×10^{-1}	10^{-1}
$\frac{1}{100}$	0.01	1×10^{-2}	10^{-2}
$\frac{1}{1000}$	0.001	1×10^{-3}	10^{-3}
$\frac{1}{10000}$	0.0001	1×10^{-4}	10^{-4}
$\frac{1}{100000}$	0.00001	1×10^{-5}	10^{-5}
$\frac{1}{1000000}$	0.000001	1×10^{-6}	10^{-6}

Standard form

Like powers of 10, a number in standard form has a similar format:

$a \times 10^n$ where $1 \leqslant a < 10$, and n is a positive or negative whole number.

For example, the distance from the Earth to the Sun, 150 000 000 000 m, is written in standard form as 1.5×10^{11} m.

The mass of a proton, 0.000 000 000 000 000 000 000 000 001 66 kg is written in standard form as 1.66×10^{-27} kg.

To convert a number bigger than 10 to standard form:

● split it into two parts – the first part comes immediately after the first non-zero integer and is a number greater than or equal to 1
● the second part is the power of 10 (that's one less than the total number of digits in the number before you would arrive at the decimal point) – this is the n in '$\times 10^n$'
● add a multiply sign between the two parts.

For example, to convert 257 000 000 to standard form:

● the decimal point comes immediately after the first non-zero integer, which is the number 2, so the first part is 2.57
● there are nine figures in the number altogether, so n = 9 − 1 = 8
● the number in standard form is 2.57×10^8.

To convert a number smaller than 1 to standard form:

● reading from left to right, write down the first digit that is not a zero and place a decimal point immediately after it
● count the number of places the decimal point has moved to the right, n
● add the '$\times 10^{-n}$' term.

For example, to convert 0.006 03 to standard form:

● the first part is the decimal that starts with the number 6; this is 6.03
● to get to the decimal point in 6.03, the point in 0.006 03 has moved three points to the right, so n = 3
● the number in standard form is 6.03×10^{-3}.

Tip

Remember:
● For numbers greater than 10, n is always positive.
● For numbers less than 1, n is always negative.
● For numbers between 1 and 10, n is zero.

(A) Worked examples

1   The number of radioactive carbon nuclei in a sample of carbon taken from a peat marsh in Cheshire in 1984 was estimated to be 530 400 000 000.

About 120 centuries earlier the number of radioactive nuclei in that sample would have been 1 060 100 000 000.

Write these numbers in standard form.

a   530 400 000 000

   Step 1 The first part is 5.304, which corresponds to 'a' and is between 1 and 10

   Step 2 There are 12 digits altogether, so n = 12 − 1 = 11

   Step 3 Write down the two parts, 5.304×10^{11} nuclei

b   1 060 100 000 000

   Step 1 The first part is 1.0601, which corresponds to 'a' and is between 1 and 10

   Step 2 There are 13 digits altogether, so n = 13 − 1 = 12

   Step 3 Write down the two parts, $1.060\,1 \times 10^{12}$ nuclei

2   Write the following wavelengths in standard form.

   a   Orange light (0.000 000 58 metres).

      Step 1 Write down 5.8

      Step 2 The dp is now after the 5 (in 5.8); the dp has moved seven places to the right (from 0.000 000 58 to 5.8)

      Step 3 The wavelength in standard form is 5.8×10^{-7} m

   b   X-rays (0.000 000 000 195 metres).

      Step 1 Write down 1.95

      Step 2 The dp is now after the 1 (in 1.95); the dp has moved 10 places to the right (from 0.000 000 000 195 to 1.95)

      Step 3 The wavelength in standard form is 1.95×10^{-10} m

(B) Guided questions

1   A spacecraft travels a distance of 4×10^8 metres to the Moon in a time of 2.592×10^5 seconds

Write both the distance and time as numbers in normal form.

Use your answers (and a calculator) to find the average speed. Give the speed in standard form.

Step 1 Convert to normal form:

distance to Moon = 4×10^8 m = 400 000 000 m

time = 2.592×10^5 s = ... s

Step 2 Divide to find speed:

speed = $\dfrac{\text{distance}}{\text{time}}$ = ... m/s

Step 3 Convert speed to standard form:

speed = ... m/s

2 There are approximately nine million atoms in every cubic metre of space near the Orion Nebula.

Find, in standard form:

a the volume in m³ of a cube of space of side 200 m near the Orion Nebula

Step 1 Find the volume of the cube:

volume = length × width × height = 200 m × 200 m × 200 m = ... m³

Step 2 Write in standard form = ... m³

b the number of atoms in that cube.

Step 1 number of atoms = volume in m³ × 9 000 000

= .... × 9 000 000

= .... atoms

Step 2 Write in standard form = ... atoms

C Practice question

3 The surface of the Earth is divided into plates. In the North Atlantic Ocean, two of these plates meet. These plates are moving apart at about 25 mm per year. How far apart will they move in five hundred thousand years? Give your answer in metres in standard form.

Using a calculator with numbers in standard form

In your exam you may need to use standard form with a scientific calculator. Most calculators have a display of around nine numbers across the screen, which means very large numbers and very small numbers cannot be entered in normal form.

If, for example, we wanted to calculate $(2.99 \times 10^3) \times (4.1 \times 10^8)$ we would:

Step 1	key in 2.99
Step 2	press the '×10ˣ' key (on some calculators the '×10ˣ' key is labelled 'EXP')
Step 3	key in 3
Step 4	key in ×
Step 5	key in 4.1
Step 6	press the '×10ˣ' key
Step 7	key in 8
Step 8	press the '=' key to show the answer, 1.2259×10^{12}

To enter 2.99×10^{-3}, press the − or ± key before entering the number 3.

Tip

On many calculators, if you press the '=' key after entering a number in standard form, the number is displayed in normal form. Unfortunately, this does not work in reverse. However, pressing the 'ENG' button shows the number in engineering form, which is similar to, but not exactly the same as, standard form.

Fractions

Normal form (decimals), standard form, fractions and percentages are all numbers that we can key into a calculator. As all of these forms represent numbers, they can be changed from one form to another, as shown in Table 1.6.

Table 1.6 Numbers in different forms

Normal form	Standard form	Fraction	Percentage
0.03	3×10^{-2}	$\dfrac{3}{100}$	3%
0.5	5×10^{-1}	$\dfrac{1}{2}$	50%
3.7	3.7×10^{0}	$\dfrac{37}{10}$	370%
12.25	1.225×10^{1}	$12\dfrac{1}{4}$	1225%

In the following sections we will look at fractions, ratios and percentages, which you will have to use in many calculations.

Fractions (the traditional way)

Most scientific calculators have the ability to give answers as a fraction, such as $\frac{5}{24}$, although you can set your calculator to display a decimal number instead. To convert from a fraction to a decimal, you need to divide the numerator (top number) by the denominator (bottom number):

$\frac{5}{24}$ 0.208 (3 dp).

Your calculator will have a button to do this calculation for you, often labelled $S \Leftrightarrow D$.

You also need to be able to multiply, divide, add and subtract fractions.

Multiplying fractions

Multiplication is very straightforward. You multiply together the numbers on the top (numerators) and then the numbers on the bottom (denominators).

For example: $\frac{2}{3} \times \frac{3}{4} = \frac{6}{12}$

We can simplify the fraction by dividing the numerator and denominator by 6 to give $\frac{1}{2}$.

Dividing fractions

To divide fractions, we invert the divisor (the second fraction) and multiply.

For example: $\frac{3}{4} \div \frac{7}{8} = \frac{3}{4} \times \frac{8}{7} = \frac{24}{28}$

We can simplify the fraction by dividing the numerator and denominator by 4 to give $\frac{6}{7}$.

Adding and subtracting fractions

It is easy to add or subtract fractions if they have the same denominator.

Example 1: $\frac{2}{7} + \frac{4}{7} = ?$

Here, the common denominator is 7, so $\frac{2}{7} + \frac{4}{7} = \frac{6}{7}$

Key terms

Numerator: The number above the line in any fraction.

Denominator: The number below the line in any fraction.

Example 2: $\frac{7}{12} - \frac{5}{12} = ?$

Here, the common denominator is 12, so $\frac{7}{12} - \frac{5}{12} = \frac{2}{12}$

We can simplify the fraction by dividing the numerator and denominator by 2 to give $\frac{1}{6}$.

This is fairly straightforward, but if we are not given a common denominator we need to find one. Any two (or more) fractions will share a common denominator that will allow us to add or subtract them. To find a common denominator you can multiply the two denominators together.

For example:

$\frac{1}{3} + \frac{1}{4} = ?$

In this example, we can't add them together as is. However, if we multiply the denominators together, we get $4 \times 3 = 12$. To make sure the fractions stay equivalent (the same) when we change the denominator, we also have to multiply the numerator by the same number. This is because $\frac{1}{3}$ is not the same value as $\frac{1}{12}$. We multiplied the denominator (3) by 4 to get 12, so we need to multiply the numerator (1) by 4 as well. If we do this to both fractions we get:

$\frac{4}{12} + \frac{3}{12} = ?$

Now we can add them together to get $\frac{7}{12}$.

Fractions (the easy way)

In every GCSE Physics exam you are expected to be able to use a scientific calculator. Learning how to use a calculator to work out fractions is straightforward.

The instructions below tell you how to enter the fraction $\frac{3}{4}$ and then add $\frac{1}{2}$.

- Look for the fraction button; the symbol will appear on the screen when you press it.
- Press the number 3 button; the 3 is displayed as the numerator.
- Press 'down' on the navigation button when you are ready to enter the denominator.
- Press the number 4 button.
- Press 'right' on the navigation button so the fraction, $\frac{3}{4}$, is on the screen.
- Press +.
- Enter the fraction $\frac{1}{2}$ in the same way as you entered $\frac{3}{4}$, finishing by pressing 'right' on the navigation button.
- Press =.
- The display should show the answer: $\frac{5}{4}$ or $1\frac{1}{4}$.
- Press S⇔D to see the answer displayed as a decimal (1.25).

Mixed fractions are fractions that have a whole number bit and a fraction bit. For example, $2\frac{1}{2}$ is a mixed fraction.

The instructions below tell you how to enter the mixed fraction $2\frac{1}{2}$ and then multiply by $3\frac{3}{4}$.

Tip

If you come across a fraction where the numerator is bigger than the denominator, such as $\frac{9}{6}$, this means that it can be simplified as a whole number and a fraction. In this example you have $\frac{6}{6}$ plus another $\frac{3}{6}$, or $1\frac{3}{6} = 1\frac{1}{2}$.

Tip

If the common denominator you get by multiplying both denominators together is very high, see if you can simplify it by reducing it to a smaller number. For example, the fraction sum $\frac{24}{56} + \frac{8}{56}$ might be easier understood as $\frac{3}{7} + \frac{1}{7}$, as both fractions share the same multiple of 8. If you divide both fractions by 8, you get a much simpler sum. You can also do this simplification after completing the sum, if you prefer.

Tip

It will be useful to look at your own calculator when following these instructions.

- Find and press the mixed fraction symbol – this is usually above the fraction button; you need to press the shift button, then the fraction button.
- You can now enter a mixed fraction.
- Press the number 2 button, and then press 'right' on the navigation button.
- Press the number 1 button, and then press 'down' on the navigation button.
- Press the number 2 button, and then press 'right' on the navigation button.

- The display now shows $2\frac{1}{2}$.
- Press ×.

- Enter the mixed fraction $3\frac{3}{4}$ in the same way as you entered $2\frac{1}{2}$.

- Press =.

- The display should show $9\frac{3}{8}$ or $\frac{75}{8}$.

- Press S⇔D to see the answer displayed as a decimal (9.375).

Once you know how to enter fractions in your calculator, you can add, subtract, multiply and divide them easily.

Tip

Remember that using a calculator is a skill and skills are not learned overnight. They need practice!

(A) Worked examples

1 **The total resistance R of resistors of $12\,\Omega$ and $6\,\Omega$ when placed in parallel, is given by this equation:**

$$\frac{1}{R} = \frac{1}{12} + \frac{1}{6}$$

Use the equation to find the value of R.

Step 1 Add the fractions, $\frac{1}{12} + \frac{1}{6}$.

Step 2 The denominators are 12 and 6.

We can't write $\frac{1}{12}$ as a fraction out of 6, but we can write $\frac{1}{6} = \frac{1\times 2}{6\times 2} = \frac{2}{12}$, so:

$$\frac{1}{R} = \frac{1}{12} + \frac{1}{6} = \frac{1}{12} + \frac{2}{12} = \frac{3}{12} = \frac{1}{4}$$

Step 3 Link back to $\frac{1}{R}$:

Since $\frac{1}{R} = \frac{1}{4}$, then $R = 4\,\Omega$

2 **Two thirds of the chemical energy in a petrol driven lawnmower is wasted as heat. One third of the remaining energy is wasted as sound. The rest is useful energy.**

What fraction of the input energy is useful?

Step 1 Find fraction *not* wasted as heat: $1 - \frac{2}{3} = \frac{3}{3} - \frac{2}{3} = \frac{1}{3}$

Step 2 Find fraction wasted as sound: $\frac{1}{3} \times \frac{1}{3} = \frac{1}{9}$

Step 3 Find total fraction wasted (heat + sound): $\frac{2}{3} + \frac{1}{9} = \frac{6}{9} + \frac{1}{9} = \frac{7}{9}$

Step 4 Subtract from 1 to find useful energy: $1 - \frac{7}{9} = \frac{9}{9} - \frac{7}{9} = \frac{2}{9}$

So, $\frac{2}{9}$ of the total input energy is converted into useful energy.

B Guided question

1 **A manufacturer claims that more than $\frac{3}{4}$ of the electrical energy input to a chainsaw is transferred to useful kinetic energy.**

A gardener uses 60 kJ of electrical energy in the chainsaw and finds that 10 kJ are wasted as heat and sound.

Is the manufacturer's claim valid?

Step 1 Calculate the useful energy: $60\,\text{kJ} - 10\,\text{kJ} = \ldots\ldots\ldots\,\text{kJ}$

Step 2 Calculate the useful fraction: $\dfrac{\ldots\ldots\ldots\text{kJ}}{60\,\text{kJ}} = \dfrac{\ldots\ldots}{6}$

Step 3 Which is bigger $\dfrac{3}{4}$ or $\dfrac{?}{6}$?

$\dfrac{3}{4} = \dfrac{\ldots\ldots}{12}$ or $\dfrac{?}{6} = \dfrac{\ldots\ldots\ldots}{\ldots\ldots}$

Step 4 Write a conclusion: The bigger fraction is $\dfrac{\ldots\ldots}{\ldots\ldots}$, so the manufacturer's claim is ....

C Practice questions

2 Complete these calculations using a calculator.

a $1\frac{1}{4} + 3\frac{5}{8}$ d $3\frac{2}{5} - 4\frac{7}{10}$ g $1\frac{2}{3} \div \frac{4}{9}$

b $2\frac{2}{3} + 4\frac{5}{6}$ e $2\frac{1}{4} \times 3\frac{5}{8}$ h $1\frac{1}{2} \div 2\frac{1}{4}$

c $7\frac{5}{12} - 6\frac{1}{4}$ f $2\frac{2}{5} \times 5\frac{5}{6}$

3 Alloys are mixtures of metals. Gold is often alloyed with copper to make it harder and more hard-wearing.

9-carat gold contains $\frac{3}{8}$ pure gold and $\frac{5}{8}$ copper by mass.

Calculate the mass of a piece of 9-carat gold if it contains 95 g of copper.

4 A marine physicist finds that a sample of seawater contains 35 g of salt and 965 g of water.

Use this data to find:

a the fraction of seawater that is common salt, giving your answer in its lowest terms

b the mass of pure water obtained in a desalination plant when 100 kg of seawater is treated.

> **Tip**
>
> You know that $\frac{5}{8}$ of the gold is copper, so use this to first find the mass of $\frac{1}{8}$, and then use this amount to find the mass of $\frac{8}{8}$ (in other words, all) of the 9-carat gold.

Ratios

A ratio is the number of times a quantity is bigger or smaller than another. For example, a step-up transformer may have a turns ratio $N_s : N_p$ of $3 : 1$, which means that there are three times as many turns on the secondary coil N_s as there are on the primary coil N_p.

> **Key term**
>
> Ratio: A way to compare quantities; for example, three apples and four oranges are in the ratio $3 : 4$.

Notice the use of the two dots (:) between the 3 and the 1; this shows it is a ratio. Sometimes it is easier simply to express a ratio as a whole number, as opposed to a fraction. In the example above we might say 'the turns ratio in the step-up transformer is 3'.

For a ratio to be valid, the quantities being compared must be **of the same unit**. So, a ratio, even when expressed as a whole number, a fraction or decimal, does not usually have a unit.

Ratios are also used to show direct proportion.

Look at the first two rows of Table 1.7. When we double (or triple, or quadruple) the mass, we do the same thing to the volume – that's what direct proportion means.

Table 1.7 $m : V$ ratios

Mass, m (g)	10	20	30	40
Volume, V (cm^3)	2	4	6	8
Ratio $m : V$	10:2 = 5:1	20:4 = 5:1	30:6 = 5:1	40:8 = 5:1

The last row in the table also shows that the ratio $m : V$ is always the same (in this case 5:1). The constant ratio is a test for direct proportion. But ratios are like fractions. In this case the ratio $m : V$ (or, if you prefer, the fraction $\frac{m}{V}$) is 5.

From the definition of density ρ you know that $\rho = \frac{m}{V}$. So, the ratio needs a unit; the unit for density is g/cm^3.

(A) Worked examples

1 **The total input energy to a simple motor is 3000 J. The useful output energy is 1.8 kJ. Calculate the motor's efficiency.**

Step 1 Write down what we mean by efficiency:

$$\text{efficiency} = \frac{\text{useful output energy}}{\text{total input energy}}$$

Step 2 Substitute the numbers: $\text{efficiency} = \frac{1800 \text{ J}}{3000 \text{ J}}$

Notice that in this question we changed the 1.8 kJ to 1800 J. This is because when we calculate a ratio, both numbers must have the same unit.

Step 3 Do the calculation: efficiency = 0.6

2 **The ratio of voltages in an ideal transformer is the same as the ratio of the number of turns on the primary and secondary coil.**

A transformer has 125 turns on the primary coil and 25 turns on the secondary coil. The voltage in the primary coil is 24 V.

a **Calculate the turns ratio $N_s : N_p$.**

Step 1 Write down what we mean by turns ratio:

$$\text{turns ratio} = \frac{\text{secondary turns}}{\text{primary turns}}$$

Step 2 Substitute the numbers: $\text{turns ratio} = \frac{25 \text{ turns}}{125 \text{ turns}}$

Step 3 Do the calculation: turns ratio = 0.2 (or 1:5 or $\frac{1}{5}$)

b **Calculate the voltage in the secondary coil.**

Step 1 Write down what we mean by voltage ratio:

$$\text{voltage ratio} = \frac{\text{secondary voltage}}{\text{primary voltage}}$$

Step 2 Substitute the numbers: $\text{voltage ratio} = \frac{\text{secondary voltage}}{24 \text{ V}}$

Step 3 Put voltage ratio equal to turns ratio and do the calculation:

$$0.2 = \frac{\text{secondary voltage}}{24 \text{ V}} \Rightarrow \text{secondary voltage} = 0.2 \times 24 = 4.8 \text{ V}$$

Tip

When finding a ratio where the two numbers are measuring a similar physical quantity (like energy) convert both to the same unit before doing the calculation.

B Guided question

1   A physicist designing a new rotary engine builds a scale model. The model is built to a scale of 1:20. A piston in the model has a length of 5.2 cm. A shaft in the production engine has a length of 1.4 m.

 a Explain what is meant by the statement that the model is built to a scale of 1:20.

 Step 1 This means that every part in the model is .... of the size of the corresponding part in the production engine.

 b Calculate the length of the piston in the production engine.

 Step 1 length of piston in production engine = ... times length in model

 Step 2 = ... × 5.2 cm = ... cm

 c Calculate the length of the shaft in the model.

 Step 1 length of shaft in model = length of shaft in production engine ÷ ....

 Step 2 = 1.4 m ÷ .... = ... m = ... cm

C Practice questions

2   An empty beaker has a mass of 24 g. A small ball has a mass of 18 g. Find the ratio of their masses giving your answer in the form a:b, where a and b are whole numbers.

3   The current flowing in a resistor is measured when the voltage across it is changed and the results are recorded in a table, like the one shown below. By calculating a suitable ratio, show that the voltage is directly proportional to the current.

Voltage, V (V)	3.2	4.0	4.8	5.6	6.4	7.2
Current, I (A)	0.20	0.25	0.30	0.35	0.40	0.45
Ratio						

4   A machine is supplied with 3000 J of energy. It wastes 480 J of this input energy. The rest is converted into useful work. By calculating a suitable ratio, find the machine's efficiency, giving your answer as a decimal to 2 dp.

Percentages

It can be difficult to compare fractions when they have different denominators. For example, it is not easy to say whether $\frac{3}{10}$ is bigger or smaller than $\frac{4}{11}$ without doing some calculations. Percentages solve that problem.

> **Key term**
>
> Percentage (%): A fraction of 100.

A percentage is a fraction of 100.

Table 1.8 Some common percentages, decimals and fractions

Fraction	$\frac{1}{20}$	$\frac{1}{10}$	$\frac{1}{4}$	$\frac{1}{2}$	$\frac{3}{4}$	1
Decimal	0.05	0.10	0.25	0.50	0.75	1.00
Percentage	$\frac{1}{20} = \frac{5}{100}$ $= 5\%$	$\frac{1}{10} = \frac{10}{100}$ $= 10\%$	$\frac{1}{4} = \frac{25}{100}$ $= 25\%$	$\frac{1}{2} = \frac{50}{100}$ $= 50\%$	$\frac{3}{4} = \frac{75}{100}$ $= 75\%$	$\frac{2}{1} = \frac{100}{100}$ $= 100\%$

(A) Worked examples

1 The amount of heat leaving a particular room is shown in this table. Find the percentage of heat leaving through the walls.

Heat lost through	Percentage heat loss
windows	35%
floor	5%
ceiling	35%
door	5%
walls	?

Step 1 The percentage heat loss though the windows, door, ceiling and floor is 35% + 5% + 35% + 5% = 80%.

Step 2 So, the percentage heat loss through the walls is 100% − 80% = 20%

2 The efficiency of a car engine is 32%. If the engine is supplied with 150 MJ, how much energy is wasted?

Step 1 32% of the energy supplied is useful, the rest is wasted.

Step 2 32% of 150 MJ = $\frac{32}{100} \times 150 = 48$ MJ

Step 3 So, wasted energy = 150 − 48 = 102 MJ

(B) Guided questions

1 Eighteen carat gold is a mixture that contains 75% pure gold. The rest is made up of other metals. A company buys a block of 18 carat gold that has a mass of 1.2 kg.

In this sample, calculate:

a the mass of pure gold

Step 1 mass of gold = ...% of 1.2 kg = $\frac{\text{.......}}{100} \times 1.2$ kg = ...kg

b the percentage of other metals

Step 1 percentage of other metals = ...% − 75% = ...%

c the mass of other metals

Step 1 mass of other metals = 1.2 kg − .... kg = ... kg

2 A helical spring is 20 mm long. In a Hooke's Law investigation, it extends elastically to a total length of 32 mm. What percentage of its original length is the spring's extension?

Step 1 Calculate extension: extension = ... mm − .... mm = ... mm

Step 2 Calculate fraction of original length:

fractional extension = $\frac{\text{......mm}}{\text{......mm}}$ = ...

Step 3 Convert to percentage: percentage extension = ... × 100% = ...%

> **Tip**
> Remember the rules for converting fractions, decimals and percentages.

C Practice questions

3 A mixture contains 80 g of oil and 120 g of water. What percentage of the mixture is oil?

4 A car is supplied with 30 MJ of chemical energy. Of this, 21 MJ are wasted and the rest is converted into useful kinetic energy.

   What percentage of the input energy is:

   a wasted
   b converted into useful energy?

5 A physicist estimates that the power available in a waterfall is 1500 kW, but only 12% of this available power could be captured by a turbo-generator and changed into electrical power. How much electrical energy could be obtained from this waterfall every minute?

Estimating results

Making an estimate is an important skill in physics. It helps you to judge quickly if an answer is roughly correct. For example, if somebody told you that the speed of an athlete is 100 m/s, a quick estimate shows you that they must be wrong as the world record for 100 m is just under 10 m/s.

★ **Not explicitly required for WJEC/Eduqas GCSE Physics.**

Estimates can be simple guesses based on your experiences, or they can be based on quick calculations.

The first step, when estimating, is to convert the numbers to 1 significant figure (sf). So, for example, instead of multiplying by 112, we would multiply by 100. Instead of dividing by π (which is roughly 3.14), we would divide by 3. Rather than divide by 19.3, we would divide by 20, and so on.

A Worked examples

1 **In 2009, Michael Phelps swam two lengths of a 50 m swimming pool in a time of 49.82 s – a world record. Estimate his average speed.**

   Step 1 average speed $= \frac{\text{distance}}{\text{time}} = \frac{(2 \times 50)}{49.82}$

   Step 2 Writing the numbers to 1 sf makes this calculation easy: $\frac{100}{50} = 2 \text{ m/s}$

2 **A piece of metal of mass 199 g has a volume of 39.8 cm^3. Estimate its density.**

   Step 1 density $= \frac{\text{mass}}{\text{volume}} = \frac{199}{39.8}$

   Step 2 Writing the numbers to 1 sf makes this calculation easy: $\frac{200}{40} = 5 \text{ g/cm}^3$

Tip

Even if a skill is not explicitly required by your exam board, you will likely cover it in Maths GCSE, so it can't hurt to refresh your memory.

B Guided question

1 **An athlete runs 10 times around a rectangular track measuring 112 m by 96 m in a time of 495 s. Estimate her average speed.**

   Step 1 Calculate distance around track: distance = 100 + ... + ... + ... = 400 m

   Step 2 Calculate distance run: total distance = ... × 400 m = ... m

   Step 3 Estimate average speed: speed $= \frac{\text{distance}}{\text{time}} = \frac{......\text{m}}{......\text{s}} = ...\text{m/s}$

C Practice questions

2   A car uses 1 litre of petrol for every 22 km of distance travelled. The car-owner travels 12 250 km every year. Estimate the volume of petrol used every year.

3   Light travels at 3.0×10^8 m/s. Estimate how long it would take to travel 400 000 km to the Moon and 400 000 km back again. Give your answer to the nearest second.

4   A factory robot can do 195 J of useful work every second of the day. Estimate how much useful work the robot could do in 10 hours. Give your answer to the nearest MJ.

Using sin and $\sin^{-1}$ keys

Students taking Edexcel International GCSE must be able to use the sin and $\sin^{-1}$ keys on the calculator to solve problems on refraction. The example below shows how.

★ **Only explicitly required for Edexcel International GCSE.**

A Worked example

The diagram shows a ray of light passing through a rectangular glass prism.

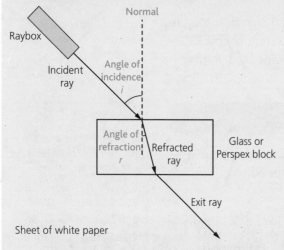

Normal

Raybox

Incident ray

Angle of incidence
i

Angle of refraction
r

Refracted ray

Glass or Perspex block

Exit ray

Sheet of white paper

Key terms

Angle of incidence, i: Angle between the incident ray and the normal to the boundary of a transparent material.

Angle of refraction, r: Angle between the refracted ray and the normal to the boundary of a transparent material.

Normal: A line drawn at right angles to a surface.

Refractive index: The ratio $\sin i : \sin r$.

a   **If the angle of incidence i in air is 40° and the angle of refraction r in glass is 25°, calculate the refractive index of the glass, n, giving your answer to 2 dp.**

Step 1 The mathematical relationship between i and r is $n = \dfrac{\sin i}{\sin r}$

Step 2 Substituting the numbers for i and r gives $n = \dfrac{\sin 40}{\sin 25}$

Step 3 To calculate n, using a calculator, do the following:

Action	Display shows
press the sin button	sin(
enter 40 and close the bracket	sin(40)
press the ÷ button	sin(40) ÷
enter sin(25) just as you entered sin(40)	sin(40) ÷ sin(25)
press =	1.52096

So, the refractive index is 1.52 (2 dp).

b   **Using your answer to part a, find the value of i when the angle of refraction is 40°, giving your answer to 2 dp.**

Step 1 To find an angle such as the value of i, we need to use the $\sin^{-1}$ key.

Step 2 Again, we will use the equation $n = \dfrac{\sin i}{\sin r}$

Step 3 If we multiply both sides of the equation by $\sin r$, we get $\sin i = n \times \sin r$.

Step 4 The first step to finding i is to calculate $\sin i$.

Step 5 So, we have to calculate $n \times \sin r$:

Action	Display shows
enter 1.52 × sin	1.52 × sin(
enter 40)	1.52 × sin(40)
press =	0.977037
So, $\sin i = 0.977037$ and we can work backwards to find i:	
Press SHIFT and then sin	$\sin^{-1}($
The calculator is looking for the value of the sin of the angle, so that it can calculate the angle.	
*press ANS	$\sin^{-1}($ANS
close the bracket and press =	77.6977

So, the angle of incidence is 77.70° (2 dp).

> **Tip**
> If your calculator does not have an ANS button, you could re-enter the number 0.977037 at the stage marked with the asterisk. Then close the bracket and press =.

B Guided question

1   **The diagram shows refraction of light when the angle of incidence in glass is equal to the** critical **angle, c. If the refractive index of the glass is 1.52, calculate the value of c, giving your answer to 1 dp.**

Step 1 $n = \dfrac{1}{\sin c} \Rightarrow c = \sin^{-1}\left(\dfrac{\cdots}{\cdots}\right)$

Step 2 So, $c = \ldots\ldots° = \ldots\ldots°$ (to 1 dp).

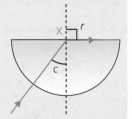

> **Key term**
> Critical angle, c: The angle of incidence in an optically dense medium when the angle of refraction in air is 90°.

C Practice questions

2   **The angle of incidence in glass that gives an angle of refraction in air of 90° is called the critical angle. Show that the refractive index of glass is 1.56 (2 dp) if the critical angle is 40°.**

3   **Using the information given in the diagram, calculate the refractive index of medium B with respect to medium A.**

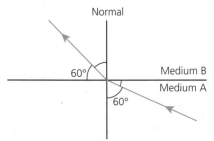

4   **Diamond has a refractive index of 2.42. Calculate the angle of refraction in diamond when the angle of incidence is 70°. Give your answer to the nearest degree.**

> **Tip**
> For advice on rearranging the subject of equations, see 'Changing the subject of an equation' on pages 42–43.

» Handling data

Using significant figures

Significant figures (sf) are approximations of a long number to give it an approximate meaning.

For example, the number 34 529 taken to:

- 5 sf is 34 529 – it stays as it is because it already has only five numbers
- 4 sf it 34 530 – the decider is the **9**, so we round up from 29 to 30
- 3 sf it 34 500 – the decider is the **2**, so we round down from 529 to 500
- 2 sf is 35 000 – the decider is the **5**, so we round up from 4500 to 5000
- 1 sf is 30 000 – the decider is the **4**, so we round down from 4529 to 0000.

When deciding on the number of significant figures, leading zeros after the decimal point do not count but trailing zeros do count.

For example, if the number 0.034 529 001 taken to:

- 5 sf is 0.034 529
- 4 sf is 0.034 53
- 3 sf is 0.0345
- 2 sf is 0.035
- 1 sf is 0.03

Giving the correct number of significant figures in calculations is important in physics because it signifies the degree of precision.

Suppose a balance reads to the nearest 100 g:

- If the true mass on the balance was larger than 2450 g but less than 2500 g, the balance would round up to the nearest 100 g and give a reading of 2500 g.
- If the true mass was bigger than 2500 g but less than 2550 g, the balance would round down to the nearest 100 g and also give a reading of 2500 g.
- For all masses between 1000 g and 9900 g, this balance would give a figure to 2 sf.
- For this balance, the last two digits will always be zero.

Suppose we look at a new balance capable of reading to the nearest 10 g and measure the same mass, and we get a reading of 2470 g.

- We would know for sure that the mass (M) can be estimated as $2465\,g \leqslant M < 2475\,g$.
- For all masses between 1000 g and 9990 g, this balance would give a figure to 3 sf.
- For this balance, only the final digit would always be zero.

Suppose this second balance gave a reading of 2500 g, like the first balance. Could we say that the two balances are giving the same information? The answer is no.

- The first balance can only tell us that $2450\,g \leqslant M < 2550\,g$, although the reading is 2500 g (2 sf).
- The second balance is telling us that $2495\,g \leqslant M < 2505\,g$, although the reading is also 2500 g (3 sf).

Key terms

Significant figures (sf): Approximations to a number, determined by a set of mathematical rules.

Leading zeros: Zeros *before* the first significant figure in small numbers; for example, there are two leading zeros (the zeros before the 2) in 0.002 034.

Trailing zeros: Zeros *after* the first significant figure; for example, the zero to the right of the 2 in 0.002 034 is a trailing zero and would be significant if expressing this number to 3 sf.

Tip

Some answers may be recurring. For example, the calculator display shows 9.652652652… If this happens, use the recurring form of the number when writing the number to so many significant figures. So, 9.652952652… is 10 (to 1 sf), 9.7 (to 2 sf), 9.65 (to 3 sf), 9.653 (to 4 sf) and so on. You might find it useful to look again at the section on decimal places on page 6 it is very similar to significant figures.

Tip

If you are doing CCEA, you are specifically required to know how to express a physical quantity to an appropriate number of significant figures, and write them to 1, 2 or 3 decimal places.

It is clear that the number of significant figures tells us something of the degree of precision in the instrument being used to measure it.

(A) Worked examples

1 A rectangular piece of metal measures 21.4 cm by 15.3 cm. Calculate its area to an appropriate number of significant figures.

Step 1 area = 21.4 cm × 15.3 cm = 327.42 cm^2

Step 2 Each number in the question was to 3 sf, and as there is no other guidance, the most appropriate answer is 327 cm^2

2 An irregular solid of mass 320 g displaces 55 cm^3 of water. Calculate the density of the solid to an appropriate number of significant figures.

Step 1 density = $\dfrac{\text{mass}}{\text{volume}} = \dfrac{320}{55} = 5.818181...$ g/cm^3

Step 2 The number given in the question to the least number of significant figures was 55 cm^3 (2 sf)

Step 3 So, the most appropriate answer is 5.8 g/cm^3 (2 sf)

> **Tip**
> When answering mathematical questions, look at the data (numbers) in the question that are given to the least number of significant figures. Your final answer should have the same number of significant figures, unless the question tells you otherwise.

(B) Guided question

1 A current of 1.4 A flows through a resistor of 6.8 Ω. Calculate the voltage across the resistor, giving your answer to an appropriate number of significant figures.

Step 1 Write equation for Ohm's Law: $V = I \times R$

Step 2 Substitute for I and R: V = ...

Step 3 Do the arithmetic: V = ... volts

Step 4 Number of sf in data in question is two.

Step 5 Give answer to appropriate number of sf: V = ... volts.

(C) Practice questions

2 How much heat energy is needed to raise the temperature of 2.55 kg of water by 12.2 °C if the specific heat capacity of water is 4200 J/kg°C? Give your answer to an appropriate number of significant figures.

3 Find, to an appropriate number of significant figures, the kinetic energy of a ball of mass 55 g moving at a speed of 19 m/s.

4 There are 6.02×10^{23} molecules in 18 g of water. Find, to an appropriate number of significant figures, the number of molecules in 1 g of water.

> **Tip**
> The scientific equations you might need to answer these questions are as follows, but see if you can remember them without looking, then check your knowledge:
> $$\Delta E = mc\Delta\theta$$
> $$E_k = \tfrac{1}{2}mv^2$$

Finding arithmetic means

There are three different types of average. But the one most commonly used by GCSE Physics students is the arithmetic mean, or simply 'the mean'.

To find the mean of a collection of numbers, we add them all up and divide the sum by the number of numbers in the collection. This helps us to get a number that is likely to be closer to the true measurement.

Suppose we were trying to find the density of a wooden rod. To find the volume we need to know the diameter of the rod. We could measure it once, but we have no assurance that the measured value is reliable. So, we might repeat the measurement five more times. We would then have six values that are all slightly different. Instead of choosing one, we take a mean (or average).

We also need to be aware that some of our diameters may be a little too big or small. By taking the mean, we hope that the numbers that are too big cancel out the numbers that are too small (sometimes known as outliers).

(A) Worked example

A student measures the time it takes for a pendulum to swing 10 times. The student uses a stopwatch capable of measuring time to two decimal places. He takes the measurement five times. The results are: 8.07 s, 7.83 s, 8.14 s, 8.23 s, 8.10 s

Find the mean time for one swing.

Step 1 Add the values together: 8.07 + 7.83 + 8.14 + 8.23 + 8.10 = 40.37

Step 2 Divide by the number of values: mean = 40.37 ÷ 5 = 8.074 s

Although the mean value of the experimental results is 8.074 s, the best value to use is 8.07 s. This is because no matter how many measurements we make with this stopwatch, we can never measure a time correctly to 3 dp.

So, the mean time for one swing = 8.07 ÷ 10 = 0.807 s ≅ 0.81 s.

Tip

If we see a figure that is an outlier in a set of data, it is best to exclude it when finding the mean.

Tip

The symbol ≅ means 'is roughly equal to'.

(B) Guided question

1   **Five students independently measure the resistance of a length of wire. They obtain these results: 2.1 Ω, 2.2 Ω, 0.5 Ω, 1.9 Ω, 1.8 Ω**

   a   **Identify the outlier.**

      Step 1 The outlier is: ....

   b   **Calculate the mean of the other four resistances.**

      Step 1 The sum of the other results is 2.1 + ... + ... + ... = ...

      Step 2 The mean resistance is .... ÷ 4 = ... Ω

(C) Practice questions

2   Five students each measure the diameter of a metal rod. Their results, in mm, are: 312, 317, 313, 314, 314

   Calculate the mean diameter to 3 sf.

3   The generally accepted value for the specific heat capacity of water is 4.2 J/g°C. A group of 10 students measure the specific heat capacity of water and their mean result is the generally accepted value.

   Nine of the students' results, in J/g°C, are: 4.1, 4.2, 4.2, 4.3, 4.3, 4.1, 4.2, 4.0, 4.1

   Calculate the value for the specific heat capacity obtained by the tenth student.

Mode and median

The other kinds of average, less frequently used in GCSE Physics, are mode and median. Mode means 'most common'. Median means 'in the middle', when arranged in increasing order.

Suppose a physics student measures, to the nearest gram, the mass of 21 marbles and sets out the results in order of increasing mass, as below.

14, 14, 14, 15, 15, 15, 15, 16, 16, 16, 16, 17, 17, 17, 17, 17, 17, 17, 17, 18, 18

In this example, the most common number in the list is 17. This means that the modal mass is 17 grams (or simply the mode is 17 g)

Now look at the highlighted number, 16. Of the 21 numbers in the list, 10 are to the left of it and 10 are to the right. The number 16 is in the middle. So, the median mass is 16 grams.

It's easy to find the number in the middle when the total number is odd. Just add 1 to the total number of results (do not add all the results together) and divide by 2. In this case, $(21 + 1) \div 2 = 11$, so the median is the 11th number counting from either end.

It's slightly trickier when there is an even number of numbers, as there are two numbers in the middle. Consider the following two lists of six numbers.

List A: 2, 3, 5, 5, 6, 7   List B: 2, 3, 4, 5, 6, 7

In List A, the middle numbers are both 5. So, the median is 5.

In List B, the middle numbers are 4 and 5. They are different, and so we take the median to be half-way between them. So, the median is 4.5.

> **Tip**
>
> For information on how to use frequency tables to find mean, mode and median see pages 28–30.

(A) Worked example

Seven pebbles have masses, in grams, of 7, 2, 3, 3, 5, 6 and 9. Find:

a **the modal mass**

Step 1 In order, the masses are: 2, 3, 3, 5, 6, 7, 9

Step 2 The modal mass is the most common, which is 3 g.

b **the median mass.**

Step 1 The median mass is in the middle, which is the fourth of the seven numbers, 5 g.

B Guided question

1 The diameters of nine ball bearings are measured to the nearest mm. The mode is 9 and the median is 6. Seven of the nine diameters, in mm, are: 4, 1, 7, 6, 2, 3, 9

The largest diameter is 9 mm. Find the missing diameters if both of them are known to be greater than 6 mm.

Step 1 Arrange the numbers in order: ....., ..., ..., ..., ..., ..., ...,

Step 2 Since the mode is 9, there must be at least .... 9s. So, add .... to the ordered list. There are now .... numbers in the ordered list.

Step 3 The median is the .... number in the ordered list, so the missing number must be ...., ..., ... or .... Since the mode is 9, the missing number cannot be .... or .... So, the missing number must be .... or ....

C Practice question

2 Twenty students carry out an experiment to find the density of ethanol. Below are their results, in g/cm^3.

0.79	0.79	0.78	0.78	0.78	0.77	0.77	0.79	0.79	0.77
0.77	0.79	0.79	0.79	0.79	0.78	0.78	0.78	0.78	0.77

Find:

a the mode
b the median density, in g/cm^3.

Constructing frequency tables, bar charts and histograms

Frequency tables

Physicists often have to collect data before it can be processed. One form of processing this data is in a frequency table. There are two types of frequency table – ungrouped and grouped.

When you want to measure frequency, but no additional order is needed, you use an ungrouped frequency table. Suppose, for example the lengths of 100 similar screws are measured to the nearest mm and their results are recorded in a table. This is an ungrouped frequency table.

Sometimes, there is so much data that it needs to be grouped into classes. For example, a physicist may be interested in the number of times the electronic components in certain pieces of apparatus fail, but it might be enough to know the whether the number of failures was less than 5, between 5 and 10, between 10 and 15 and so on. This would require a grouped frequency table.

Both types of table are used to obtain statistical information, such as the mean, mode and the median.

★ **Edexcel International students are not explicitly required to construct and interpret frequency tables.**

Key terms

Ungrouped frequency table: A table of data in which each item is discrete.

Grouped frequency table: A table of data in which items are ranges in classes or groups.

Constructing frequency tables

When constructing frequency tables, you need to draw a table with the key data in the left-hand column and a tally column next to it.

In the following example, each time the diameter of a ball bearing is measured, a tally mark is placed in the appropriate part of the table. The first four tally marks are vertical, but the fifth tally mark is a diagonal across the previous four, so that the tallies are bundled into groups of five. This makes it easier to count quickly – you can total the frequency in the right-hand column. Table 1.9 is an example of an ungrouped frequency table. In Table 1.10, there are four classes for the masses of the ball bearings. This is a grouped frequency table.

Table 1.9 Example of an ungrouped frequency table

Diameter, D (mm)	Tally	Frequency
21	卌 卌 I	11
22	卌 II	7
23	IIII	4
24	卌 III	8
	Total	30

Table 1.10 Example of a grouped frequency table

Mass, m (g)	Tally	Frequency
$10 \leqslant D < 15$	卌 卌 卌 I	16
$15 \leqslant D < 20$	卌 卌 卌 卌 II	22
$20 \leqslant D < 25$	卌 卌 IIII	14
$25 \leqslant D < 30$	卌 III	8
	Total	60

A Worked example

A physics student measures the horizontal range R, in cm, of 40 marbles when they are rolled horizontally over the edge of a bench. The results are shown below.

Using groups of 0–5 cm, 5–10 cm, and so on, construct a grouped frequency table by means of a tally chart.

3	4	16	17	15	5	9	10	19	15
1	7	16	17	12	6	7	11	17	10
19	1	14	10	7	4	1	11	13	16
7	8	1	11	17	17	9	2	10	2

Range R (cm)	Tally	Frequency
$0 \leqslant R < 5$	卌 IIII	9
$5 \leqslant R < 10$	卌 IIII	9
$10 \leqslant R < 15$	卌 卌	10
$15 \leqslant R < 20$	卌 卌 II	12
	Total	40

Tip

To avoid multiple entries, cross off data items (or highlight/circle them with a coloured pen) as you construct the tally chart.

C Practice question

1 A physics student wants to use dice to simulate radioactive decay. For the experiment to be valid, the dice must not be biased. This means that each one of the six numbers must be equally likely to turn up when the dice are thrown.

The student throws a single die 60 times and obtains the results shown below. The student is told that the die is probably unsuitable for the experiment if any number comes up fewer than eight times, or more than 12 times.

3	5	1	6	1	1	5	5	5	6	1	5
2	2	6	2	4	4	6	4	3	3	6	5
3	2	4	6	1	3	4	3	6	2	6	4
1	2	4	4	1	4	3	3	4	2	1	6
1	3	2	4	2	3	6	5	4	3	5	3

a Construct an ungrouped frequency table to represent the data.
b Use your table to decide whether or not the die is biased.

Using ungrouped frequency tables to find mean, mode and median

You can also use ungrouped frequency tables to quickly find mean, mode and median. For example, suppose the lengths of 100 similar engine components are measured to the nearest mm, and their results are recorded in a table. This is an ungrouped frequency table.

Length (mm)	98	99	100	101	102	**Total**
Frequency (number of items)	4	15	66	12	3	**100**

From this table we can calculate the mean, mode and median as follows.

Mean

We could set out the sum as 98 + 98 + 98 + 98 + 99 + 99 + 99 + (and so on).

However, there is a much quicker way. Multiply the frequency by the length, and add the totals together.

$$\text{mean length} = \frac{(4 \times 98) + (15 \times 99) + (66 \times 100) + (12 \times 101) + (3 \times 102)}{100}$$
$$= 99.95\,\text{mm}.$$

Mode

The number 66 is the highest frequency in the table. This shows there are 66 lengths of 100 mm. The mode is, therefore, 100 mm.

Median

Work out which number is in the middle when the items are arranged in order. In this case, there are 100 items and they are already in order of length. Since the number of items is even, there are two values in the middle. These are the 50th and the 51st values. The first 19 items are accounted for by the two lowest lengths. The next table entry shows 66 items each of length 100. So, the 50th and 51st items are to be found here. The median length is, therefore, 100 mm.

Using grouped frequency tables to find mean, mode and median

Grouped frequency tables can be used in a similar way. For example, suppose a physicist measures the mass of 40 different ball bearings (to the nearest gram) and presents her results as a grouped frequency table.

Class (g)	30–34	35–39	40–44	45–49	50–54	Total
Frequency	4	10	15	6	5	40

Mean

To calculate an estimate of the mean from this data, we imagine that, on average, the mass of each item in the class is the mid-point value (sometimes called the *class mark*). We can, therefore, add another row to the table.

Class (g)	30–34	35–39	40–44	45–49	50–54	Total
Frequency	4	10	15	6	5	40
Class mark (g)	32	37	42	47	52	

We can use the class mark (mid-point mass) to estimate the mean as we did with the ungrouped data.

$$\text{estimate of mean mass} = \frac{(4 \times 32) + (10 \times 37) + (15 \times 42) + (6 \times 47) + (5 \times 52)}{40}$$

$$= 41.75\,\text{g}$$

Mode

The highest frequency in the table is 15, so the modal class is 40–44. The modal length within that class cannot be determined by GCSE level Maths, so we will ignore this.

Median

Since there are 40 items, we need to find the 20th and 21st items to find the median. There are 14 items with mass between 30 and 39 grams. The next table entry shows 15 items with masses between 40 and 44 grams. So, the 20th and 21st items are both within the class 40–44. The median class is, therefore, 40–44 grams.

Bar charts

Bar charts are simple graphs used to show discrete data. For example, a physics student carried out a survey into the main type of heating used in 100 different households and recorded the results in the table below.

Type of central heating	Coal	Oil	Gas	Wind	Geothermal	Total
Number of households	4	15	66	12	3	100

We can use this data to draw a bar chart.

Key terms

Bar charts: Charts showing discrete data in which the height of the unconnected bars represents the frequency.

Discrete data: Data that can only have particular values, such as the number of marbles in a jar.

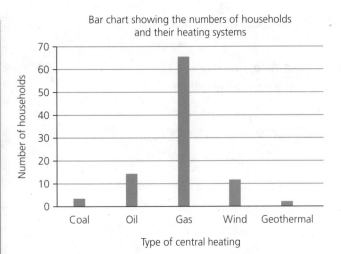

Bar chart showing the numbers of households and their heating systems

Each bar tells us the number of households using a particular type of heating. The longer the bar, the greater the number of households. Notice also:

- the axes are labelled exactly as in the table
- the bars are not joined together
- each bar has a label at the bottom.

Bar charts can also be used for continuous data.

(A) Worked example

This table shows the lengths of 500 springs, to the nearest mm, received from a manufacturer.

Length of spring (mm)	98	99	100	101	102	Total
Number of springs	20		330	60	15	500

One of the entries in the table is missing. Calculate the number of 99 mm springs in the sample and then use the data to draw a bar chart.

Step 1 The number of springs accounted for in the table is 20 + 330 + 60 + 15 = 425

Step 2 The number of 99 mm springs is the total number of springs − 425 = 500 − 425 = 75

Step 3 The bar chart is drawn on a grid (or lined paper) with a vertical scale going up to 350 (because the maximum bar height is 330 mm).

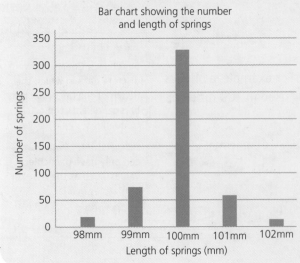

Bar chart showing the number and length of springs

B Guided question

Tip
This question shows how you can also place two bars side-by-side.

1 Every year since 2014, the physics department in a sixth form college has recorded the numbers of male and female students taking GCSE Physics. The results from 2018 to 2014 are shown in the bar chart.

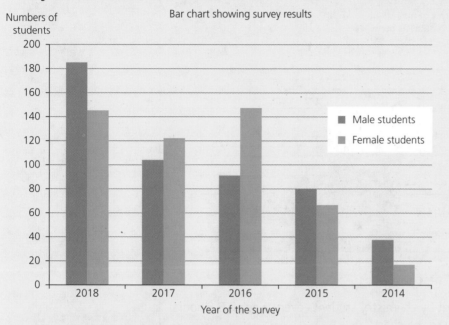

Bar chart showing survey results

Use the chart to find:

a the year in which the number of female students is greatest

Step 1 Female students have a ....-coloured bar.

Step 2 This bar is biggest in ....

b the year in which the difference between the numbers of male students and female students was greatest

Step 1 Year in which difference between males and females is greatest is the year in which there is the greatest difference between the .... of the bars.

Step 2 This year is ....

c the total number of students taking GCSE Physics in 2015.

Step 1 Number of male students in 2015 = ...

Step 2 Number of female students in 2015 = ...

Step 3 Total number of students in 2015 = ... + ... = ...

C Practice question

2 Below is an incomplete bar chart of a survey carried out to find the average time spent by students studying biology, chemistry, mathematics and physics every week in a school.

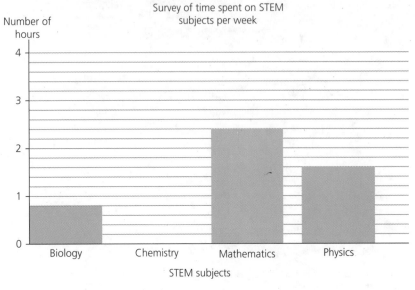

Survey of time spent on STEM subjects per week

Number of hours

STEM subjects

a The number of hours spent by chemistry students every week is 3.3 hours. Add the bar for chemistry to the chart.

b How much more time is spent by physics students than biology students?

c What is the total time spent by the average student studying these 4 subjects?

Histograms

Histograms are similar to bar charts but are only used for continuous data, such as the length or mass of components. This means that in a histogram the bars **must** touch.

The histogram below shows the percentage mark gained by a group of students in a class. Notice that the percentage mark is a continuous variable, and the bars are all of equal width (5 percentage marks).

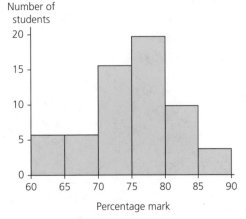

Histogram showing distribution of marks

Number of students

Percentage mark

Key terms

Histograms: Charts showing continuous data in which the area of the bar represents the frequency.

Continuous data: Data that can take any value within a range, such as the mass of a beaker.

Tip

The histograms shown in this book are all of equal width as histograms with bars of unequal width are beyond the scope of GCSE Physics.

A Worked example

A physicist measured the wavelengths, to the nearest 10 cm, at which empty lemonade bottles resonated. She recorded the results in a table, as shown. Show these data in a histogram.

Wavelength λ (cm)	150	160	170	180	190	200	210	220	230
Number of bottles	2	3	4	5	6	7	8	6	5

Step 1 Since the wavelengths are measured to the nearest 10 cm, the 150 cm wavelength covers $145\,cm \leqslant \lambda < 155\,cm$, the 160 cm wavelength covers $155\,cm \leqslant \lambda < 165\,cm$, and so on.

Step 2 These limits are the lower and upper boundaries of the wavelengths. They allow us to draw this grouped frequency table.

Wavelength λ (cm)	Class limits	Number of bottles
150	$145\,cm \leqslant \lambda < 155\,cm$	2
160	$155\,cm \leqslant \lambda < 165\,cm$	3
170	$165\,cm \leqslant \lambda < 175\,cm$	4
180	$175\,cm \leqslant \lambda < 185\,cm$	5
190	$185\,cm \leqslant \lambda < 195\,cm$	6
200	$195\,cm \leqslant \lambda < 205\,cm$	7
210	$205\,cm \leqslant \lambda < 215\,cm$	8
220	$215\,cm \leqslant \lambda < 225\,cm$	6
230	$225\,cm \leqslant \lambda < 235\,cm$	5

Step 3 In this case, the width of each bar is always 10 cm as can be seen from the class limits.

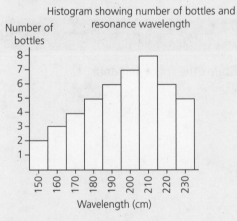

Histogram showing number of bottles and resonance wavelength

Step 4 The wavelength corresponding to 150 cm is a bar stretching from 145 cm to 155 cm, which are the class limits. The adjacent bar stretches from 155 to 165 cm, and so on. In this way the bars touch – because the data are continuous.

1 The amount of snow that fell over a 20 day period in a winter resort was recorded. The results are shown in this table.

Snow fall S (mm) (class)	Class mid-point Snow fall (mm)	Number of days
$10 \leqslant S < 20$	15	3
$20 \leqslant S < 30$		6
$30 \leqslant S < 40$		
$40 \leqslant S < 50$		3
$50 \leqslant S < 60$		2

Show this data on a histogram.

Step 1 The number of days has to add up to 20.

So, the number of days when the snowfall was between 30 and 40 mm was ....

Step 2 The class mid-point is exactly half way between the maximum and minimum snowfall.

So, the numbers missing from the middle column of the table are ...., ..., ... and ....

Step 3 Draw a set of axes. The vertical axis is labelled *Number of days*. The horizontal axis is labelled .... and will range from 0 to ....

Step 4 The first bar is centred on 15 mm, ... days in height and .... mm wide. Draw this bar.

Step 5 Draw the remaining bars. The final bar is two days in height, centred on .... mm and .... mm wide.

Step 6 Add the title to the histogram.

C **Practice question**

2 A student sorts a box of springs according to their lengths and obtains the following data.

CLASS	
Spring length (mm)	Number of springs
Greater than 20 but less than 30	1
Greater than 30 but less than 40	3
Greater than 40 but less than 50	7
Greater than 50 but less than 60	9
Greater than 60 but less than 70	12
Greater than 70 but less than 80	7
Greater than 80 but less than 90	3
Greater than 90 but less than 100	1

a Show these data on a histogram.
b How many springs are there altogether?
c Which group of springs contains the median length of the springs?
d Which group contains the modal length of springs?

Pie charts

Pie charts display the proportions of a whole data set as sector angles or sector areas. The total angle around the centre of the pie is 360° and the area of the sector is proportional to the angle at the centre.

Suppose a physics student investigated the way in which 60 different students came to school. The data collected was to be shown in a pie chart. We can use this information to work out the following:

60 students = 360°

So, 1 student = 360 ÷ 60 = 6°

This means that, for each sector, the angle at the centre of the pie chart would be calculated using:

angle = number of students × 6°

★ **Construction and interpretation of pie charts are only explicitly required for CCEA GCSE Physics.**

Worked example

This pie chart represents the energy resources of a European country.

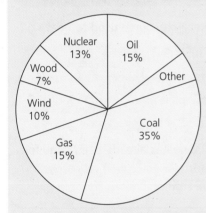

a **What percentage of the energy resource is represented by 'Other'?**

Step 1 Total percentages given = 35% + 15% + 10% + 7% + 13% + 15%
= 95%

Step 2 So, 'Other' represents 100% − 95% = 5%

b **What is the sector angle of the Wind sector?**

Step 1 The Wind sector represents 10% of the pie.

Step 2 So, its sector angle is 10% of 360° $\equiv \dfrac{10}{100} \times 360 = 36°$

c **The sector labelled Gas represents an energy of 8.1×10^{17} J.**

Step 1 Calculate the total energy resources used by this country.

Step 2 Gas represents 15% $\equiv 8.1 \times 10^{17}$ J

Step 3 1% $\equiv \dfrac{8.1 \times 10^{17}}{15} = 5.4 \times 10^{16}$

Step 4 100% $\equiv 100 \times (5.4 \times 10^{16}) = 5.4 \times 10^{18}$ J

B Guided question

1   The results of a survey into the method used by pupils to come to school showed the following:

Method	Walking	Cycling	Car	Bus	Train
Number of pupils	15	5	35	20	15

Complete the table and show the results in a pie chart.

Step 1 The number of students surveyed altogether was: ....

Step 2 Each student in the pie chart is represented by an angle of .... degrees.

Step 3 So, the angles for each method of transport are:

Walking = 60°; Cycling = ... °; Car = ... °; Bus = ... °; Train = ... °

Step 4 With a compass, draw a large circle to represent the pie.

Step 5 With a ruler, draw a line from the centre of the circle to its circumference.

Step 6 Draw the sectors in the pie using the angles found in Step 3, then label the sectors.

C Practice question

2   Below are the results of a survey of 180 people asked about the main energy resource used to heat their homes.

Energy resource	Gas	Oil	Coal	Electricity	Wood	Other
Number of people	90	45	25	10		2
Sector angle (degrees)			50			

a   Calculate the number of people whose main energy resource is wood.

b   Calculate the sector angle for each resource, if displayed as a pie chart. One has already been done for you.

c   Show the data on a pie chart.

Simple probability

A probability is a number which expresses the likelihood that a particular event will occur. A probability of 1 means the event is certain to occur. A probability of 0 means the event is certain not to occur. All probabilities must lie between 0 and 1.

The probability of an event, E, occurring is defined as the ratio:

$$\text{probability of E occuring} = \frac{\text{total numbered of favourable outcomes}}{\text{total number of possible outcomes}}$$

In GCSE Physics you are most likely to meet probability when studying radioactive decay.

Suppose that the probability that any individual nucleus will decay in a given minute is 0.2, and that initially we have a population of 1000 undecayed nuclei.

After 1 minute, 200 would be expected to decay, leaving 800 undecayed.

★ **Only explicitly required for GCSE students studying WJEC or Edexcel International Physics.**

Tip

Phrases such as 'Probability of E occurring' are very wordy. Physicists tend to shorten it to P(E).

After a further minute, we would expect 160 more to decay, leaving 640 nuclei undecayed.

This process would continue as shown:

Time elapsed (mins)	0	1	2	3	4	5	6	7
Expected number of undecayed nuclei	1000	800	640	512	410	328	262	210

A quick look at the table shows that the time taken for the number of undecayed nuclei to fall to half of its original number is a little over 3 minutes. Physicists recognise this as the half-life for that decay.

A Worked examples

Tip

Although we might know the probability that a particular nucleus will decay in a given time interval, we cannot say for certain whether or not it will decay. The process is random and spontaneous. This means that the experimental graphs for decay are likely to be more ragged than those you see in textbooks or examination papers.

1 **What is the probability of drawing an ace from a pack of 52 playing cards when the pack is cut once?**

Step 1 number of favourable outcomes = 4 (there are 4 aces in the pack)

Step 2 total number of possible outcomes = 52 (there are 52 cards in the pack altogether)

Step 3 probability of an ace = $\frac{4}{52} = \frac{1}{13}$.

2 **These are the results of a survey of the main energy used by 180 people to heat their homes:**

Energy resource	Gas	Oil	Coal	Electricity	Wood	Other
Number of people	90	45	25	10	8	2

Calculate the probability that a person chosen at random from the sample:

a **uses oil**

Step 1 P(uses oil) = $\dfrac{\text{number of people who use oil}}{\text{total number of people in survey}}$

$= \dfrac{45}{180}$

$= \dfrac{1}{4}$

b **does not use coal.**

Step 1 P(does not use coal) = $\dfrac{\text{number of people who do not use coal}}{\text{total number of people in survey}}$

$= \dfrac{(180 - 25)}{180}$

$= \dfrac{155}{180}$

$= 0.861$

B Guided question

1 **This graph shows a decay curve for a radioisotope.**

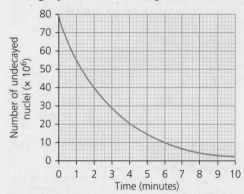

a **Find the half-life of the isotope starting with the original number of undecayed nuclei.**

Step 1 In one half-life, the number of undecayed nuclei falls by .... %

Step 2 So, in one half-life the 80 million undecayed nuclei will fall to .... million.

Step 3 From the graph this takes .... minutes.

b **Use your answer to part a to state the probability that a particular nucleus in the sample is likely to decay within a period of 6 minutes.**

Step 1 In 6 minutes, the number of undecayed nuclei has fallen to .... million.

Step 2 So, the fraction that have decayed is ....

Step 3 So, the probability of decay within 6 minutes is ....

C Practice question

2 Suppose a sample contains 3000 undecayed nuclei and the probability a particular nucleus will decay in a given minute is 0.3.

a Copy and complete the table below to show the expected number of undecayed nuclei every minute up to 7 minutes. One entry has been completed for you.

Time elapsed (mins)	0	1	2	3	4	5	6	7
Expected number of undecayed nuclei	3000					504		

b Plot the graph of *Expected number of undecayed nuclei* (*y*-axis) against *Time elapsed in minutes* (*x*-axis).

c Use your graph to show that the half-life of this decay is approximately 1.9 minutes.

Using a scatter diagram to identify a correlation

★ **Not explicitly required for WJEC or CCEA GCSE Physics.**

Correlation is about finding out if there is any relationship between two variables, such as the temperature on a summer day and the number of ice-creams sold at the seaside.

Suppose a market researcher carries out a survey by counting the number of people who buy an ice cream (X) from a particular store between 10.00 am and 2.00 pm every day over a period of 31 days. Every day the market researcher also measures the average temperature (Y). A graph of X against Y is called a scatter graph. The connection between the variables is sometimes called the correlation. For example, we might expect the number of ice creams bought would be greater on those days when it is hotter.

There are three types of correlation: positive correlation, negative correlation and no correlation.

> **Tip**
> •
> In Edexcel International GCSE, the terms 'pattern' and 'trend' are used instead of 'correlation'; if you are studying this specification, treat these words as interchangeable.

Positive correlation means that, as one variable (X) increases, the other variable (Y) also increases.	
Negative correlation means that as, one variable (X) increases, the other variable (Y) decreases.	
If there is **no connection** between the variables, we say there is no correlation.	

Examples of positive correlation are distance travelled in a given time and speed, length of a spring and the applied force.

Examples of negative correlation are current and resistance, frequency and wavelength of electromagnetic waves.

Key terms

Scatter graph: A graph plotted between two quantities to see if there might be a relationship between them.

Positive correlation: This occurs if one quantity tends to increase when the other quantity increases.

Negative correlation: This occurs if one quantity tends to decrease when the other quantity increases.

No correlation: There is no relationship whatever between two quantities.

Tip
• • • • • • • • • • • •
To identify correlation, ask yourself what happens to one variable when the other increases.
- If one goes up when the other goes up, it is positive correlation.
- If one goes up when the other goes down, it is negative correlation.
- If there is no relationship, it is no correlation.

Two points to note:

1 Do not think that, because there is positive correlation, there is a **causal relationship**. If you were to investigate the number of trees growing in a person's garden, you might find a correlation with the number on their front door. But that is not to say the number of trees increases just because the number on the door increases.

2 On the other hand, a positive correlation might also show a causal relationship. For many years, tobacco companies said that there was a correlation between smoking and lung cancer, but there was no causal relationship and it was just coincidence. Now on every pack of cigarettes there is a clear message: Smoking Kills.

> **Key term**
>
> Causal relationship: The reason why one quantity is increasing (or decreasing) is that the other quantity is also increasing (or decreasing).

A Worked example

In an experiment, a physicist measures the temperature T at different distances, D from a light source. Plot the data as a scatter graph with T on the vertical axis and d on the horizontal axis. State the correlation, if any, between d and T and describe what it means.

Distance d (cm)	1	2	3	4	5	6	7
Temperature T (°C)	30	29	27	24	21	21	20

Step 1 The scatter graph shown has a trend line of negative slope. The correlation is, therefore, negative.

Step 2 We interpret this as meaning that the temperature decreases as the distance from the light source increases.

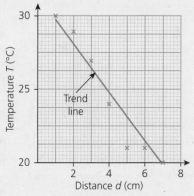

B Guided question

1 **A teacher wanted to find out if students performed best when the weather was hot.**

He recorded the average examination scores of his pupils and the average outside temperature when the test was taken, and recorded the results in a scatter graph.

State the type of correlation shown by the scatter graph and give a reason for your answer.

Step 1 Look to see if there is a connection between the points and decide on the correlation:

....

Step 2 Give a valid reason for your answer:

....

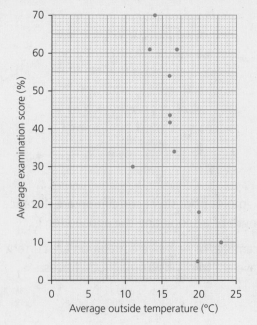

C Practice question

2 In a computer model of the molecules of a gas, each molecule changes its direction randomly when it collides with another molecule. The time between the collisions is also random. The distance a particular molecule is from a fixed point therefore changes with each collision.

A small data set is generated from the computer model and is shown in this table.

Distance d of molecule from a fixed point (mm)	0	15	21	26	30	33	38	40
Collision number N	0	1	2	3	4	5	6	7
$\sqrt{N}$	0		1.4		2.0			

It has been suggested that there is some correlation between the distance from the fixed point and the square root of the collision number N.

Complete the table, entering the numbers to one decimal place, and then plot a scatter graph of *Distance d* against $\sqrt{N}$.

What type of correlation, if any, is suggested by the scatter graph?

Tip

When answering questions about correlation, you may find it helpful to draw a line or curve of best fit.

Making order of magnitude calculations

★ **Not explicitly required for CCEA GCSE Physics.**

Suppose you estimate the floor area of the room you are in to be $80\,\text{m}^2$. A physicist would say that the order of magnitude of the floor area is $100\,\text{m}^2$ or $10^2\,\text{m}^2$.

If you express any number in standard form (see page 9), then the nearest power of 10 represents the order of magnitude.

For example, the mass of the Sun is approximately $2 \times 10^{30}\,\text{kg}$, so its order of magnitude is $10^{30}\,\text{kg}$.

The mass of the Earth is known to be $6 \times 10^{24}\,\text{kg}$, but since 6×10^{24} is closer to 1×10^{25} than 1×10^{24}, the mass of the Earth has an order of magnitude of $10^{25}\,\text{kg}$.

Standard form is important because physicists measure things that are incredibly small, such as the diameter of a proton (with an order of magnitude of $10^{-15}\,\text{m}$), and things that are incredibly large, such as the size of the observable Universe (with an order of magnitude of $10^{27}\,\text{m}$).

Key term

Order of magnitude: If we write a number in standard form, the nearest power of 10 is its order of magnitude.

A Worked example

A prototype rocket engine can produce a thrust of 600 kN. If the mass of the space rocket is 4500 kg, calculate its acceleration. Give your answer as an order of magnitude.

Step 1 force = mass × acceleration

Step 2 $600\,000 = 4500 \times$ acceleration

Step 3 acceleration $= \dfrac{600\,000}{4500} = 133\,\text{m/s}^2 = 1.33 \times 10^2\,\text{m/s}^2$

Step 4 The order of magnitude of the acceleration is $10^2\,\text{m/s}^2$.

B Guided question

1 **In the hydrogen atom an electron orbits a proton. The gravitational force on the electron is 4.06×10^{-47} N. The electrical force on the proton is 9.22×10^{-8} N.**

Calculate how many times greater the electrical force is than the gravitational force. Give your answer as an order of magnitude.

Step 1 electrical force ÷ gravitational force = ...N ÷ ....N = ...

Step 2 So, as an order of magnitude, the electrical force is .... times more than the gravitational force.

C Practice questions

2 What is the order of magnitude of:

  a the mass of an adult human
  b the number of seconds in a day?

3 According to an article in a national newspaper, the diameter of a sub-atomic particle called a quark is of the order of 10^{-18} m. Write down a number, in standard form, that might represent this diameter.

4 The age of the universe is 13.8 billion years. Write this as an order of magnitude.

> **Tip**
> When doing order of magnitude calculations, you often have to enter numbers into your calculator in standard form. It is worthwhile referring back to the section on page 11 to revise how to do that before attempting such questions.

» Algebra

Algebra is a branch of mathematics that uses equations in which letters represent numbers.

Physics students need to know how to solve various equations by:

- re-arranging the equation to change the subject – this step is only required if the letter (or value) on its own side is not what you want to calculate
- substituting in the correct numbers for each letter or value
- calculating the answer.

Using mathematical algebraic symbols

For algebraic calculations, you are expected to recognise and be able to use the usual mathematical symbols ($+$, $-$, $\times$, $\div$, $\sqrt{}$, and so on) but you also need to know the meaning and use of some additional special symbols, as shown.

> ★ **Not explicitly required for WJEC GCSE Physics.**

Table 1.11 Mathematical symbols

Symbol	Meaning	Example	
$<>$ or $\neq$	not equal to	$6 <> 4$	6 is not equal to 4
		$6 \neq 4$	
$>$	greater than	$6 > 4$	6 is greater than 4
$\propto$	directly proportional to	$F \propto x$	Force is directly proportional to extension
$\sim$	approximately equal to	$\pi \sim 3.14$	Pi is approximately equal to 3.14

> **Tip**
> You will see that the symbol $\propto$ appears again when looking at graphs (see pages 47–55).

Changing the subject of an equation

A lot of the equations in GCSE Physics only have three variables.

For most three-variable equations, it is possible to use the 'magic triangle' to change the subject of an equation.

For example:

- To make W the subject, cover up the W with your thumb to reveal $W = m \times g$
- To make m the subject. cover up the m with your thumb to reveal $m = \dfrac{W}{g}$
- To make g the subject, cover up the g with your thumb to reveal $g = \dfrac{W}{m}$

Magic triangles can be useful, but it is much better to develop the mathematical skills you need to solve equations without them. That way you can apply these transferable skills to solve a much wider range of problems.

If you have more than three variables, or you do not want to use a magic triangle, keep things simple. Remember:

- Take one step at a time.
- What you do to one side of an equation, you must also do to the other.

In other words, if you multiply one side of an equation by a value (whether the value is a number or a letter) you need to do the same to the other side.

For example, if you want z to be the subject in this equation:

$$x = y + z$$

… you need to subtract y from both sides:

$$x - y = +z \equiv z = x - y$$

If you want y to be the subject in this equation:

$$x = y \times z$$

… you need to divide by z on both sides. Dividing by z on the right-hand side will cancel out the multiplying by z:

$$x \div z = y \equiv y = x \div z$$

Similarly, if you wanted the subject to be z in this equation, you would divide by y on both sides.

Worked examples

One of the equations of motion is $v^2 = u^2 + 2as$.

Rearrange this equation to make s the subject.

Step 1 Work out what the question is asking: at the moment, v^2 is the subject, because the equation starts $v^2 = \dots$ We have to change this to be $s = \dots$

Step 2 Subtract u^2 from both sides: $v^2 - u^2 = u^2 + 2as - u^2$

Step 3 Simplify the right-hand side (RHS): $v^2 - u^2 = 2as$

Step 4 Divide both sides by $2a$: $\dfrac{v^2 - u^2}{2a} = \dfrac{2as}{2a}$

Step 5 Simplify the RHS: $\dfrac{v^2 - u^2}{2a} = s$

Step 6 Switch RHS and LHS: $s = \dfrac{v^2 - u^2}{2a}$

B Guided question

1   **The energy, *E*, stored in a stretched spring is given by the equation** $E = \frac{1}{2}kx^2$.

   Rearrange the equation to make *x* the subject.

   Step 1 Multiply both sides by 2: $2E = \ldots\ldots\ldots$

   Step 2 Divide both sides by k: $\frac{2E}{k} = \dfrac{}{k}$

   Step 3 Simplify: $\frac{2E}{k} = \ldots\ldots\ldots$

   Step 4 Take the square root of both sides: $\sqrt{\frac{2E}{k}} = \ldots\ldots\ldots$

   Step 5 Switch RHS and LHS: $x = \ldots\ldots\ldots$

> **Tip**
> For squaring (^{2}) and square rooting ($\sqrt{\ }$) you also need to perform the same action to both sides, just as you would for any other mathematical function (adding, subtracting, dividing or multiplying).

Substituting values into an equation

Once you have amended the subject of the equation to whatever you want to calculate (if necessary), the next step is to replace the letters with any values you have.

> **Tip**
> You may find it easier to substitute values into an equation *before* changing the subject. See what works best for you.

A Worked example

The power of an electric drill is 250 W. If the drill is switched on for 40 seconds, how much work is done?

Step 1 State the equation (with the correct subject): $W = P \times t$

Step 2 Substitute in the values for power and time: $W = 250 \times 40$

Step 3 Carry out the calculation: $W = 10\,000$ J

B Guided question

1   **A car accelerates from rest to 30 m/s in 12 s. Find its acceleration.**

   Step 1 State the equation: $a = \ldots\ldots\ldots\ldots$

   Step 2 Substitute in the values: $a = \ldots\ldots\ldots\ldots$

   Step 3 Carry out the calculation: $a = \ldots\ldots\ldots\ldots$ m/s^2

> **Tip**
> The most common mistake in substituting values is to substitute a value for the wrong letter. Take your time to check which letter stands for which value. If your final answer seems wrong or unrealistic, go back to see if you have made this mistake.

Solving simple equations

Solving an equation means carrying out the final calculation. If you use a calculator, it is unlikely you will get this calculation wrong. However, you might still make a mistake rearranging or substituting, so make sure you take care with all previous steps before solving.

Remember that, in mathematical questions, examiners generally look for a formula, correct substitutions, a correctly calculated answer and – if required – a unit. So it is a good idea to substitute values for letters as soon as possible. If you try to rearrange the letters in the equation and get it wrong, it could cost you the substitution, arithmetic and final answer marks.

A Worked example

Find the speed of a golf ball of mass 45 g if its kinetic energy is 7.29 J.

There are three challenges in this question.

- The unit for mass in the KE equation is in kg, not g.
- The formula we know is for kinetic energy, not speed.
- The formula involves v^2, rather than v.

Step 1 Convert the mass from g to kg: $m = 45\,g \equiv \dfrac{45}{1000} = 0.045\,kg$

Step 2 State the formula for kinetic energy: $E_k = \dfrac{1}{2}mv^2$

Step 3 Substitute for E_k and m: $7.29 = \dfrac{1}{2} \times 0.045 \times v^2$

Step 4 Simplify: $7.29 = 0.0225 \times v^2$

Step 5 Divide both sides by 0.0225: $\dfrac{7.29}{0.0225} = \dfrac{0.0225 \times v^2}{0.0225}$

Step 6 Simplify: $324 = v^2$

Step 7 Take square roots: $\sqrt{324} = v$

Step 8 Simplify: $v = 18\,m/s$

B Guided questions

1 **The moment of a force of 6 N about a point is 15 N cm.**

   How far away from the point does the force act?

   Step 1 Write down the relevant equation: $M = Fd$

   Step 2 Substitute letters for values: .... = ... $\times d$

   Step 3 Do the arithmetic to find d and add the unit: $d =$ ....

2 **A small experimental submarine can withstand water pressure of 1370 kN/m² before its hull begins to buckle. The density of seawater is 1050 kg/m³.**

   Calculate the depth at which the hull would buckle. Assume g = 10 N/kg.

   Step 1 Write down the relevant equation: $P =$ ....

   Step 2 Substitute letters for values (remember that pressure is in kN/m²):

   .... = ... $\times$ .... $\times$ ....

   Step 3 Do the arithmetic on the RHS: .... = ... $\times h$

   Step 4 Divide to find h: .... = ┄┄┄ = ... m
   ┄┄┄

> **Tip**
> Highlight anything in a mathematical question that is unusual, like an unexpected unit, as it will be there for a reason. It often means you have to do a further step if you are to get the right answer.

C Practice question

3 The speed of orange light in air is 3×10^{8} m/s and its frequency is 5×10^{14} Hz. Find its wavelength, giving your answer in standard index form.

4 The extension of a spring is 25 mm when the applied force is 7.5 N. Calculate the spring constant in N/cm.

5 Find the average speed of a car in m/s if it completes a journey of 54 km in 1 hour.

6 The specific heat capacity of water is 4.2 J/g°C. A bottle contains 2.5 kg of water. Overnight the temperature of the water falls from 95 °C to 25 °C. Calculate the amount of heat energy given out by the water as it cools.

7 A bag of sand falls with zero initial speed from a hot air balloon at a height of 250 m. The bag falls with an acceleration of 10 m/s². Calculate the speed of the bag when it hits the ground.

Inverse proportion

★ **Only explicitly required for CCEA GCSE Physics.**

Direct and inverse proportion can appear in algebraic equations. Remember that inverse proportion occurs when doubling one quantity causes the other quantity to halve.

For example, the equation for pressure is $P = \dfrac{F}{A}$. If the force F remains constant, then doubling the area A will cause the pressure P to halve.

Here are some values of pressure and area:

Pressure (N/m²)	120	60	40	30	20
Area (m²)	1	2	3	4	6

You can see that the pressure is decreasing while the area is increasing. This is the first hint that there is inverse proportion. The second hint is that, when we double the area, the pressure halves. However, the conclusive test is to check the product (which is what we get when we multiply pressure and area together). In this case, the product is always 120, so we can say that the pressure is inversely proportional to the area.

Key term

Inversely proportional: Quantities x and y are inversely proportional to each other if their product xy is constant.

A Worked example

The resistance of five wires is measured. All are made of the same material and have the same length, but they have a different cross-section area. The results are shown in the table.

Resistance R (Ω)	60	30	20	15	10
Area A (mm²)	0.5	1.0	1.5	2.0	3.0

a **Show that the resistance is inversely proportional to the cross-section area of the wire.**

Step 1 Find the product, RA, for each wire:

Resistance R (Ω)	60	30	20	15	10
Area A (mm²)	0.5	1.0	1.5	2.0	3.0
RA (Ωmm²)	30	30	30	30	30

Step 2 Since the product RA is constant, we can confirm that the resistance is inversely proportional to the cross-section area of the wire.

b **Calculate the resistance when the area is 2.5 mm².**

Step 1 Use the information we know about the product, RA: $R \times A = 30$

Step 2 Substitute for A: $R \times 2.5 = 30$

Step 3 Divide both sides by 2.5 and solve: $R = \dfrac{30}{2.5} = 12\,\Omega$

Note, it may be tempting to look at the table and think that, since 2.5 is half-way between 2.0 and 3.0, the resistance should be half-way between 15 and 10 Ω (12.5 Ω). This would be incorrect.

B Guided question

1 **The power P of the electric element in a domestic iron is inversely proportional to its resistance R when voltage is constant. When the resistance is 48 Ω, the power is 1200 W.**

Calculate the power if the element is replaced with a 60 Ω element.

Step 1 Since Power P is inversely proportional to R, then $PR = a$ ....

Step 2 In this case, with the 48 Ω element, $PR = $ .... $\times$ .... $= $ ....

Step 3 With the 60 Ω element, $57\,600 = P \times$ ....

Step 4 Solve: $P = \dfrac{57\,600}{\text{..........}} = $ ....W

C Practice question

2 The intensity I of the light received from a lighthouse is inversely proportional to the square of its distance from the observer d^2. An experiment for a particular lighthouse gives the following data:

Intensity (W/m²)	720	360	240
d^2 (m²)	2	4	6

The intensity of the light received by a ship is 0.001 W/m².

a Use the table to find the value of d^2 for this ship.
b Use your answer to part a to show that the distance between the ship and the lighthouse is 1200 m.

»» Graphs

In physics exams, students are often asked to plot and interpret graphs arising from experimental data. Most of these will result in straight line graphs of positive gradient.

There are two types of straight line graphs of positive gradient, each with a general equation:

- $y = mx$, where gradient is m and the line passes through the origin (0,0); this shows proportionality,
- $y = mx + c$, where gradient is m and the line passes through the y-axis at a point (0,c); this shows a linear relationship but one that isn't proportional.

Tip

Students often think of graphs as having a straight line of best fit and a positive gradient, but remember that graphs can have a negative gradient, and that a line of best fit may be a curve.

Translating between graphical and numerical form

You may be asked to obtain numerical data from a graph in your exam. To do this, draw construction lines on the graph from the axes to meet the graph as shown in the worked example. Use the lines to read off the value on the axes.

A Worked example

This graph shows how the current in a metal wire changes as the voltage across it increases.

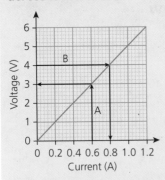

a **Use the graph to find the voltage across the wire when the current flowing through it is 0.6A.**
Step 1 Draw a vertical line (labelled A) from the point on the horizontal axis where the current is 0.6A, up to the gradient.
Step 2 Draw a horizontal line across from the point where line A crosses the gradient to the vertical axis. The reading on the vertical axis, 3V, is the answer.

b **Use the graph to find the current in the wire when the voltage across it is 4V.**

Step 1 Draw a horizontal line (labelled B) from the point on the vertical axis where the voltage is 4V, across to the gradient.

Step 2 Draw a vertical line down from the point where line B crosses the gradient to the horizontal axis. The reading on the horizontal axis, 0.8A, is the answer.

B **Guided question**

1 **This graph shows how the speed of a cyclist changes with time.**

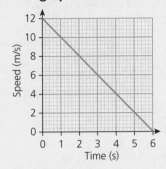

a **At what time is the cyclist travelling at 7 m/s?**

Step 1 The speed on the vertical axis is half-way between .... m/s and .... m/s.

Step 2 At this speed, draw a .... line to the graph.

Step 3 From the point where this line meets the graph, draw a .... line to the time axis.

Step 4 The line meets the time axis at .... seconds. This is the answer.

b **What is the speed of the cyclist when the time is 1.5 s?**

Step 1 The time on the horizontal axis is half-way between .... s and .... s.

Step 2 At this speed, draw a .... line to the graph.

Step 3 From the point where this line meets the graph, draw a horizontal line to the .... axis.

Step 4 The line meets the speed axis at .... m/s. This is the answer.

C **Practice question**

2 This graph shows how the total length of a spring changes with the applied force.

Find:
a the length of the spring when no force is applied
b the force needed to stretch the spring to a total length of 70 mm.

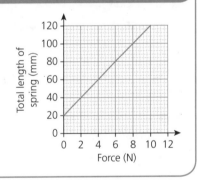

Understanding linear relationships and proportionality

As we have already seen, all straight line graphs can be written in the form $y = mx + c$. This shows a linear relationship where the graph of y against x is a straight line that does not go through the (0,0) origin. Straight lines that do go through the (0,0) origin have the form $y = mx$. These lines are special because they show direct proportion.

> **Key term**
>
> Directly proportional: Quantities x and y are directly proportional to each other if their ratio $y : x$ is constant.

A Worked example

Look at these graphs.

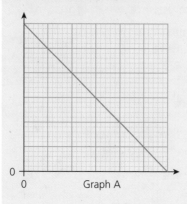

Graph A

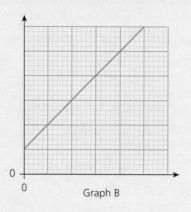

Graph B

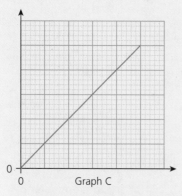

Graph C

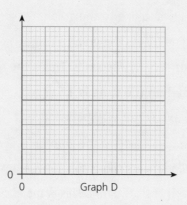

Graph D

State which of the graphs show:

a   **a linear relationship**

b   **direct proportion.**

Give reasons for your answers.

Step 1 All the graphs are straight lines – so they **all** show linear relationships.

Step 2 Only graph C is a straight line through the point (0,0), so only graph C shows direct proportion.

Step 3 Graphs A, B and D do not show direct proportion because they do not pass through (0,0).

Plotting data on a graph

Experiments usually give us a data set of two variables, *x* and *y*. Plotting a graph involves representing this data on a grid; it is one of the first steps when investigating relationships.

To plot data you need to use axes with a suitable scale that allows you to see all the points in a sensible space.

Once you have your axes and scale, you use the graph grid to help you mark with a cross (×) the position of each data point by reading up from the horizontal axes and across from the vertical axes to see where the two meet.

(A) Worked example

An experiment obtains the following data for variables x and y.

x	0	1	2	3	4	5
y	0	5	10	15	20	25

Plot, on graph paper, the graph of y (vertical axis) against x (horizontal axis) for this data.

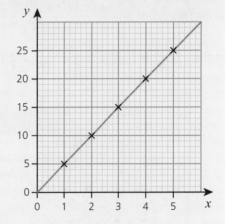

(B) Guided question

1   An experiment obtains the following data for variables x and y.

x	0	1	2	3	4	5
y	4.5	6.0	7.5	9.0	10.5	12.0

Plot the graph of y (vertical axis) against x (horizontal axis) using this data.

Step 1 Draw and label the vertical axis with the letter .... and horizontal axis with the letter ....

Step 2 Decide on the scale. The scale must be linear and cover at least half the grid.

For the y-axis, the grid is 12 cm high, so each 1 cm distance represents .... unit(s).

For the x-axis, the grid is 12 cm, so each 1 cm distance represents .... unit(s).

Step 3 The first point is at the intersection where the vertical line at $x = 0$ meets the horizontal line at $y = 4.5$. The second point is at the intersection where the vertical line at $x = 1$ meets the horizontal line at $y = ...$

Step 4 Repeat until all points are plotted.

(C) Practice question

2   Plot the graph of y against x using the data in the table.

Draw a smooth curve through the points.

x	0	1	2	3	4
y	0	1	4	9	16

Determining slope and intercept of a straight line

The equation of a graph showing a linear relationship is $y = mx + c$ and the equation of a graph showing direct proportion is $y = mx$. In these equations, m represents the slope (gradient) of the line, and c represents the intercept on the y-axis. In your exam you may need to identify both the gradient and the intercept.

Determining the intercept is usually fairly straightforward, as shown by the graph below. In this example the intercept is 20.

Calculating the slope requires a bit more work. To calculate the slope, use the formula:

$$\text{slope (gradient)} = \frac{\text{rise (total height of the gradient)}}{\text{run (total length of the gradient)}}$$

In the example below, this is: $\frac{100}{10} = 10$.

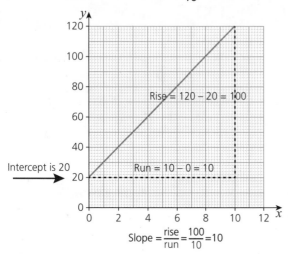

Slope $= \frac{\text{rise}}{\text{run}} = \frac{100}{10} = 10$

In some graph questions, you may also be asked to calculate the area below the line. This is the area between the graph and the x-axis. In this example you need to calculate the area of the triangle and add the area of the rectangle below the dotted line created by drawing the triangle.

To calculate the area of the triangle: $\frac{1}{2} \times \text{height} \times \text{length}$

In this example it is: $\frac{1}{2} \times 100 \times 10 = 500$.

To calculate the area of the rectangle below the dotted line: length × height

In this example it is: $10 \times 20 = 200$

By adding these totals together you get: $500 + 200 = 700$.

We will look at how to find the slopes and area for curves later.

(A) Worked example

A student measures the height H of a helical spring when compressive loads W are applied to it. They obtain the results shown below. The student is investigating the hypothesis that the compression X of the spring is related to the load W by the equation $X = kW$, where k is a constant.

W (N)	0	1	2	3	4	5
H (mm)	100	78	62	40	20	20
X (mm)			38			

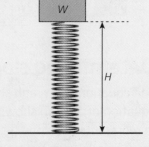

a **Complete the row for the compression X. One entry has already been done for you.**

Step 1 The length of the spring when no load is applied is 100 mm. So, the compression X is 100 mm minus the height H. This is confirmed by the data given when the load is 2 N.

Step 2 Complete the table.

W (N)	0	1	2	3	4	5
H (mm)	100	78	62	40	20	20
X (mm)	0	22	38	60	80	80

b **Plot the graph of X against W and draw the line(s) of best fit.**

Step 1 Rule the vertical and horizontal axes and label them.

Step 2 Select your scale to cover as much of the graph paper as possible, but still be easy to interpret. Always try to cover at least half of each axis with the tabulated numbers.

Step 3 Plot the points.

Step 4 There are two distinct regions – so you need to draw two straight lines with a ruler.

Step 5 Ignore any points well away from the line of best fit (outliers or anomalies). If there are points close to, but not on the line-of-best-fit, try to draw your line passing through as many points as possible, with as many points just above the line as just below it.

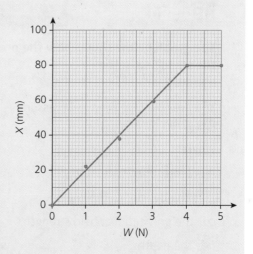

c **From your graph, find the value of k and state its unit.**

Step 1 We know that the equation of any straight line through the origin (0,0) has the form $y = mx$, where m is the gradient.

The first part of our graph is also a straight line through (0,0). It is following a similar equation, $X = kW$, where k is the gradient.

Step 2 To find the gradient we draw a large triangle on the line and calculate the ratio rise:run: $k = \dfrac{80\,\text{mm}}{4\,\text{N}} = 20\,\text{mm/N}$

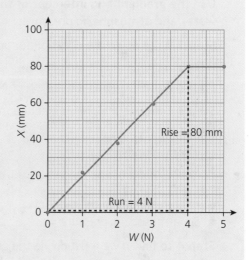

d **Explain why $X = kW$ does not apply over the entire range of the loads applied.**

Step 1 State where the equation $X = kW$ does not apply: it does not apply for weights above 4 N.

Step 2 Give a reason: beyond 4 N, the data no longer lie on the same straight line. This is because, beyond 4 N, the individual coils in the spring are touching each other –they cannot get any closer together.

Important:

The graph is a straight line through the origin for forces up to 4 N.

A straight-line graph through the origin always indicates direct proportion.

Mathematically, we write $X \propto W$.

In this case, the compression is directly proportional to the force from 0 mm to 80 mm.

B Guided question

Tip

Labels and units on graphs must be exactly the same as given in the table. As different exam boards use different forms you should always follow the exam board's example.

1 **The graph below shows how the pressure changes with depth below the Dead Sea in the Middle East.**

You are told that the total pressure P is given by $P = \rho g h + A$

where

A is atmospheric pressure

ρ is the density of the water

g is the gravitational field strength

h is the depth.

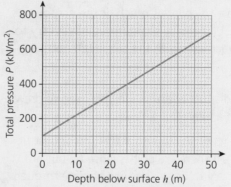

Use the gradient and intercept of the graph to find the density of the water and atmospheric pressure.

Step 1 Write down the equation for pressure P given above:
$P = $ ....

Step 2 Write the general equation of a straight line: $y = $ ....

Step 3 Compare the pressure equation and the general equation of a straight line:

- y corresponds to ....
- x corresponds to ....
- c corresponds to ....
- m corresponds to the product ....
- So, atmospheric pressure = ... kN/m^2

Step 4 To find ρ we must divide the gradient by ....

Step 5 Calculate the gradient (remember: 1 kN/m^3 = 1000 N/m^3)

$$\text{gradient} = \frac{\text{rise}}{\text{run}}$$

$$= \frac{(......- 100)\,\text{kN/m}^2}{(......- 0)\,\text{m}}$$

$$= ....\,\text{kN/m}^3$$

$$= ....\,\text{N/m}^3$$

Step 6 $\rho g = $ .... and, since g = 10 N/kg, $\rho \times 10 = $ ....

Step 7 So, the density of the water is $\dfrac{..........}{10} = $ .... kg/m^3

C Practice question

2 In an experiment, the speed of a trolley is measured at different times as it
   runs down a ramp.
   The results are shown in the table.

Speed (m/s)	1.0	1.6	2.7	2.9	3.4
Time (s)	1	2	3	4	5

a Plot the graph of *speed* (*y*-axis) against *time* (*x*-axis) and identify the
   anomalous result (outlier).
b Draw the straight line of best fit.
c Find the gradient of the graph and state its intercept on the vertical axis.
d State the equation linking the final speed *v*, the initial speed *u*, the time *t*,
   and the acceleration *a*.
e By comparing the equation for the final speed with the general equation
   of a straight line, state the initial speed of the trolley and its acceleration.

Determining slope and intercept of a curve

★ Not explicitly required in OCR or CCEA Physics.

When you plot data on a grid you must draw an appropriate line of best fit. At
GCSE this will usually be a straight line. But, occasionally, you may have data
which means your plots lie on a curve. You must draw a curve through as many
points as you can.

The worked examples show how you can find the gradient of a curve and the
area under a curve.

A Worked example

**Using the curve shown, find the rate of change of speed when the
speed is 7.5 m/s.**

Step 1 Find the point on the graph where the speed is 7.5 m/s: (3.4, 7.5)

Step 2 Use a ruler to draw the tangent at (3.4, 7.5); this is the *point of
contact* and is the only point at which the tangent touches the curve.

Step 3 The gradient of the tangent is the rate of change at the point of
contact; in this case, the gradient is 4.3 m/s^2.

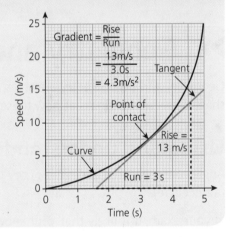

Determining the area of a curve

★ Not explicitly required in CCEA GCSE Physics.

We have seen that the gradient of a graph often has significance in physics. In some
graphs, the area between the graph and the horizontal axis is of greater interest.

For example:

● the area between speed-time and the horizontal axis represents the distance
  travelled
● the area between the force-extension graph and the horizontal axis
  represents the work done.

If the graph is a straight line, we can divide the area into triangles and squares
to find the area. To calculate the area under a curve we have to count squares.

(A) Worked example

The speed–time graph for a train is shown below.

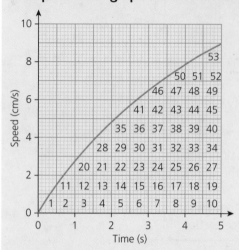

Estimate the distance travelled by the train in 5 seconds.

Step 1 The area between a speed–time graph and the time axis represents the distance travelled.

Step 2 To find this area, we count squares.

Step 3 Where the area is less than half a square, it is ignored. Where the area is more than half a square, it is regarded as a full square.

Step 4 Each full square represents a distance travelled of 1 m/s × 0.5 s = 0.5 m.

Step 5 So, the 53 squares under this curve represent a distance of 53 × 0.5 = 26.5 m.

≫ Geometry and trigonometry

Geometry and trigonometry are areas of maths that look at angles, lines and shapes. This section looks at how these skills might be required in a physics exam.

Using and understanding angles

Angles are measured in degrees. They can be any value between 0° and 360° (a full circle).

In your exam, it is likely that your knowledge of angles will be assessed in questions involving reflection and refraction. However, some exam boards may ask you to use a protractor to calculate a value:

You should know some of the more common values of angles in case you need to apply them to maths questions. For example:

- 90° is a right-angle
- 180° is a semi-circle
- 360° is a full circle – the angles of a segment in a pie chart therefore add up to 360° (see page 35)
- all angles in a triangle add up to 180°.

A Worked examples

1 The diagram shows rays of light as they pass from ethanol into the air.

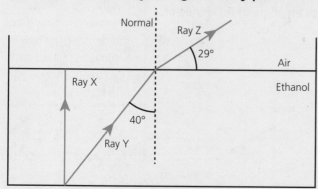

a Write down the angles of incidence of Ray X and Ray Y in ethanol.

Step 1 Ray X strikes the surface normally (at 90°), so the angle of incidence is 0°.

Step 2 Ray Y's angle of incidence is 40° (the angle between the incident ray and the normal).

b Calculate the angle of refraction of Ray Z in air.

Step 1 The angle of refraction is the angle between the refracted ray and the normal.

Step 2 This is 90° − 29° = 61°

2 A ray of red light passes through a triangular glass prism as shown.

Calculate the angle of incidence in glass at surface B.

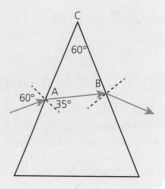

Step 1 At surface A, the angle between the ray AB and the internal surface of the glass is 90° − 35° = 55°.

Step 2 The angles in triangle ABC add up to 180°. So, at surface B, the angle between the ray AB and the glass is 180° − (55° + 60°) = 65°.

Step 3 The angle required lies between the ray AB and the normal at B. This angle is (90° − 65°) = 25°.

B Guided question

1 **Two mirrors are at right angles to each other. A ray of light falls incident on Mirror 1. The angle of incidence is 40°. The light eventually reflects off Mirror 2 as shown in the diagram.**

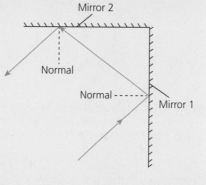

Calculate the angle of reflection **at Mirror 2.**

Step 1 Mark, on the diagram, the angles of incidence and reflection at Mirror 1, in degrees.

Step 2 Calculate the angle between the reflected ray at Mirror 1 and the mirror itself. Write the angle on the diagram.

Step 3 Calculate the angle between the incident ray on Mirror 2 and the mirror itself. (Hint: look at the triangle.)

Step 4 Calculate the angle of reflection at Mirror 2: ....

C Practice question

2 Two mirrors are inclined so that their reflecting surfaces are at 120° to each other. A ray of light falls incident on Mirror 1 with an angle of incidence of 70°. The light eventually reflects off Mirror 2.

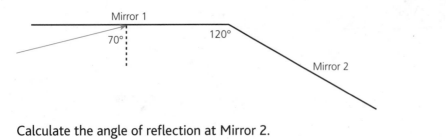

Calculate the angle of reflection at Mirror 2.

Tip

With questions on light you need a good diagram. If the examiner does not provide one, draw it yourself, and be sure to use a ruler.

2D and 3D diagrams

It is unlikely that you will be asked mathematical questions on this topic in the exam, but you do need to be able to represent 3-dimensional (3D) objects as 2-dimensional (2D) drawings.

For example, you need to know the symbols for cells, batteries, switches and lamps, and not draw the components when creating a circuit diagram.

You also need to be able to draw simple diagrams of experimental apparatus, and so on.

★ **Not explicitly required in WJEC or CCEA Physics.**

Surface area and volume

It is unlikely you will have to calculate the surface area or volume of anything other than a cube. To calculate the total surface area of a cube you need to calculate the area of each side and then multiply it by the number of faces.

If a cube has sides that are x cm long, the area of each side will be x cm $\times x$ cm, or x^2.

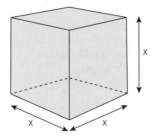

A cube has six faces, therefore, its total surface area is $6x^2$.

To calculate the volume of a cube, you need to use the formula:
length × breadth × height

As all the sides are the same length in a cube, if the side of a cube = x cm, this can also be represented = x^3.

A Worked example

A physicist is employed by a food company that makes potato chips in the form of cubes.

The bigger the surface area of the cube, the greater the amount of cooking fat it absorbs (and, therefore, the more it costs to make). The manufacturer wants to produce the greatest volume of cubes as cheaply as possible.

Should the manufacturer fry small potato cubes or large potato cubes?

Consider the possibilities for frying $64\,000\,cm^3$ of potato cubes.

Step 1 Consider the numbers of potato cubes that the manufacturer could experiment with; for example: 64 000 cubes each of side 1 cm, 8000 cubes each of side 2 cm, and 1000 cubes each of side 4 cm.

Step 2 Calculate the total surface areas:

64 000 cubes each of side 1 cm would have a total surface area of $64\,000 \times 1 \times 6 = 384\,000\,cm^2$

8000 cubes each of side 2 cm would have a total surface area of $8000 \times 4 \times 6 = 192\,000\,cm^2$

1000 cubes each of side 4 cm would have a total surface area of $1000 \times 16 \times 6 = 96\,000\,cm^2$

Step 3 Give a conclusion: To minimise surface area, the manufacturer should make chips using large cubes.

B Guided question

1   The rate at which heat can be generated by a mammal is proportional to the volume of its body, but the rate at which it *loses* heat is proportional to its surface area. By representing the mammal as a cube of side L, show that the ratio of volume:surface area is directly proportional to L (and hence mammals of bigger body size stand a greater chance of surviving low temperatures).

   Step 1 volume of cube of size L = ...

   Step 2 surface area of one face = ...

   Step 3 total surface area of cube = ...

   Step 4 volume:surface area ratio = $\dfrac{........}{........} = \dfrac{L}{6}$, which is directly proportional to L.

C Practice question

2   A metal cuboid measuring 1 cm × 2 cm × 3 cm is heated to a high temperature and then allowed to cool. By measuring the rate at which it cools, a physicist finds that at a certain temperature the cuboid is radiating energy at a rate of 88 J/s. Calculate the average rate at which each cm^2 of its surface is radiating energy.

2 Literacy

Some questions on your Physics GCSE exam papers will be extended response questions, which are usually worth six marks. As well as assessing your physics knowledge, these questions require you to construct a longer-form answer with a clear logical structure. In other words, these questions also assess the quality of your written communication (QWC), sometimes known as quality of extended response (QER).

When answering extended response questions, you need to:

- read the question carefully
- plan your response, paying particular attention to what the examiner is asking you to do
- think carefully about the time you will spend writing your answer – as a rule of thumb, you should spend no more than about a minute for every mark, so you should plan to complete a six-mark question within six minutes
- look also at the space provided for your answer – this tells you the maximum amount of the paper the examiners think you need
- be concise and to the point
- watch your spelling, punctuation and grammar.

Tip
Look at recent papers for your examination board to ensure you can recognise the extended response questions.

» How to write extended responses

Students are often concerned about extended writing questions, but there is no need to be worried. Read through this section to help you understand how to answer them effectively.

The first thing you need to do when answering one of these questions is to work out what the question is asking. Consider the following question:

Our knowledge of atomic structure is largely due to the work of Rutherford and Bohr in the early part of the 20th century.

Describe the structure of an atom, identify its constituent particles and state their locations and their properties.

In your answer you need not consider the diameters of the particles.

You can see that this question, like many six-mark questions, almost has three parts. These are:

- **Preamble**: This is the material in the first sentence, telling us that the question is about atomic structure.
- **Command line**: This is where you're told what to do – the command words are underlined.
- **Advice**: The last line is advice – a reminder that the answer should focus on structure, not size.

You may find it useful to underline or circle key words or information to help you understand the question. You can then use this information to start planning your answer.

How to plan your answer

Planning is important, as otherwise you may be tempted to write down everything that you know about the topic, even if it's not really relevant and doesn't answer the question. A quick plan can make all the difference.

First, think about the topic and decide what knowledge you need to include to answer the question. You will also have to think about how you will organise this information and set out your answer in a logical way. The next step is, therefore, to work out in what order to write this information.

For example, in the example question on the previous page, you are asked to underline(describe), underline(identify) and underline(state) information on atomic structure. You could plan your answer to this in various ways. Some possible options are using a table, a mind-map or a bulleted list. Whichever method you use, don't spend too long on your plan. The main thing is to note down the information in a way that can be ordered easily.

To answer this example question you will need to recall the names of the particles, where you find them in an atom, and their mass and charge.

Here is how you might use a table to help you organise your thoughts. As you can see, notes can be short and in draft form:

	Mass	Charge	Location
Proton	1	+1	nucleus
Neutron	1	0	nucleus
Elec	0? 1/840?	−1	orbits

The next step would be to work out a sensible order for these points. Then you are ready to write your answer.

A possible answer is:

An atom consists of a central, positively charged nucleus containing protons and neutrons. This is surrounded by electrons which orbit the nucleus in shells.

The relative charges of the proton, neutron and electron are +1, 0 and −1 respectively.

The relative masses of the proton, neutron and electron are 1, 1 and 1/1840 respectively.

The nucleus contains most of the atom's mass.

The electron shells are arranged at fixed distances from the nucleus.

How to check your answer

The final step in answering an extended literacy question is to read over your answer. This is to make sure that:

- it makes sense
- all parts of the question have been answered
- the spelling, punctuation and grammar are accurate.

Tip

Look for the command words in every question, regardless of how many marks it is worth. See pages 103–112 for more on command words and what they mean.

Tip

The examiner will mark everything you've written. If you don't want the examiner to mark your plans, remember to cross them out neatly or complete them on a separate scrap piece of paper.

Correct spelling of key scientific words is important. There are lots of technical terms in physics. You should learn the spelling of ones given below, which are commonly misspelt.

acceleration	current	meter (the instrument)
angle	displacement	metre (the unit of length)
burette	fission	potential
centre	fusion	resistance
colour	gases	rheostat
convection	insulation	temperature
coulomb	longitudinal	transmission

How to do well in extended response questions

Exam boards tend to use a banded mark scheme for extended response questions. It is made up of a level and 'indicative points' (in other words, the content required to get the marks). If you understand how extended response questions are marked, this will help you understand how you can do well in your own answers. Examiners will mark extended response answers by level and by the content contained within it.

The level of your answer will be awarded from 1–3, based on how your answer as a whole compared to the level descriptors provided. Level 1 is a weak answer, while Level 3 is a strong answer. These level descriptors instruct the examiner to look at aspects such as structure, logic, and how well the question is answered.

Once the level is identified, the examiner has to decide what mark to give the question within this band. This mark is mostly based on the indicative content. Put simply, the more indicative points you cover, the higher the mark you will achieve. You do not have to write down all of these points – for example, there may be 9 or 10 indicative points and six of these might take you into the top band.

> **Tip**
>
> You do not have to have all the indicative content to obtain full marks, but you won't get full marks if there is any incorrect information.

» How to answer different command words

Work through the following extended writing questions, which look at the main extended response command words to help you further understand how to write a good longer-form answer.

For each command word there is:

- an **expert commentary** question, which gives a sample student response along with an analysis of what is good and bad about it
- a **peer assessment** question, where you will then be given the chance to apply what you've learnt to mark a sample answer yourself
- an **improve the answer** question, where you will be asked to better another student's response in an attempt to get full marks.

Extended responses: Describe

A question requiring you to 'describe' something is asking for a detailed written account of the relevant facts and features relating to the topic being examined. Remember that describe does not mean explain, which is a higher level command word. You do not need to focus on causes and reasons in describe questions.

(A) Expert commentary

1   **Describe, in detail, how you would carry out an experiment to investigate how the resistance of a wire depends on its length. Your description should include details:**

- **of the circuit you need to set up**
- **about what you would do**
- **about what results you would record**
- **of how you would use your results to draw a conclusion.**   **[6]**

Student answer

Get a piece of wire. Connect it in a circuit with an ampmeter and voltmeter. Measure the amps and the volts. Divide the amps by the volts to get the resistence. Then repeat with another length of wire and so on until you have done it for six lengths. Then plot a graph. It will be a straight line. This tells us that the resistence is proportional to the wire.

This piece of work would probably score one mark only for the idea that the experiment requires the student to find voltage and current in order to measure resistance.

There are also spelling mistakes in two scientific terms – the correct word is ammeter (and later the student spells 'resistence' incorrectly; it should be resistance).

This is a mistake in describing how 'resistence' is calculated – we need to divide the voltage by the current, not the other way around.

Finally, the student states the resistance is proportional to the wire. However, the question makes it clear that the experiment is to show how the resistance of the wire depends on its length.

An examiner looking at this piece of work would immediately be surprised by its length. The question is worth six marks, but the student has only written a few lines of text and there is no detailed description.

It is unclear what circuit is to be constructed here.

It is unclear how the results are to be recorded.

The student is unclear what is plotted against what in the graph.

(B) Peer assessment

2   **Describe how you would measure the density of a liquid. In your answer you must:**

- identify the apparatus you would use
- state the measurements you would make
- state the equation you would use to find the density
- state one procedural precaution you would take during the experiment to obtain an accurate result
- state one safety precaution that should be taken appropriate to the experiment.   **[6]**

Tip
These bullet points are the scaffolding around which the candidate must construct an answer.

Student answer

1 Set a beaker on a top-pan balance and record its weight.
2 Fill a burette with the liquid and open the tap until the level reaches $0\,cm^3$.
3 Place the beaker and balance below the burette and add $10\,cm^3$ of liquid from the burette.
4 Record the combined weight of the beaker and liquid.
5 Repeat steps 3 and 4 until there are $100\,cm^3$ liquid in the burette.
6 As a safety precaution tie back hair and put stools and bags under the bench.

Use the mark scheme and indicative content to award this answer a level and a mark.

Mark scheme

Level descriptors	Mark
Level 3: Detailed description of a method that will work. The description is substantial and logically structured. At least 6 of the points in the indicative content are covered and the spelling, punctuation and grammar are largely accurate. The form and style are of a high standard and specialist terms are used appropriately.	5–6
Level 2: The method may lack detail, but it would work if minor changes to the procedure were made. At least 4 of the points in the indicative content are covered and the spelling, punctuation and grammar are usually accurate. The form and style are of a satisfactory standard and they have made use of some specialist terms.	3–4
Level 1: The description is quite basic and may be very short. The procedure lacks detail and would only work if major changes were made. The candidate seldom uses appropriate scientific terminology and there may be significant irrelevant and incorrect information. At least 2 of the points in the indicative content are covered. The candidate uses limited scientific terminology and there are inaccuracies in spelling, punctuation and grammar.	1–2
No relevant content	0

Indicative Content:
- measuring cylinder / pipette / burette
- linked to volume
- balance / scale(s) / top-pan balance
- linked to mass [Note: weight ≠ mass]
- subtract mass of container from combined mass of container + liquid / tare the balance
- density = mass / volume
- precaution for reliability/accuracy: avoid splashing / read to bottom of the meniscus
- safety precaution: wear safety spectacles / goggles

I would give this a level of.......... and a mark of ....

This is because ...

....

....

....

Tip

Some words in physics sound the same but have two different meanings so you need to be careful. For example, a meter is an instrument used in measurement (as in ammeter and voltmeter), but metre is a unit of length equal to $100\,cm$. Don't get them mixed up.

C Improve the answer

3 **Describe fully what determines the colour of an object. In your answer you should describe the colour of a red object in light of different colours.** [6]

Student answer

The colour of an object is determined by the colour of the light being shone on it.

Red objects absorb all the colours of the spectrum except red light. Colours that are not absorbed are reflected. So, red objects reflect red light.

Rewrite this answer to improve it and obtain the full six marks.

> **Tip**
> Try to avoid the use of bullet points in extended answer questions if you can. Using numbered points is a better option as it allows you to refer to processes more easily.

Extended responses: Explain

The command word 'Explain' means that the answer must contain some element of reasoning or justification. This justification can often be simply written, but may also contain or be entirely comprised of mathematical reasoning.

A Expert commentary

1 **Explain how a step-up transformer works. In your explanation you must:**

- **refer to the construction of the transformer**
- **refer to the type of current used in such a transformer**
- **explain the purpose of the core**
- **state the turns-ratio equation**
- **discuss the process of electromagnetic induction.** [6]

> **Tip**
> If you are to gain full marks in a question like this keep looking back at the bullet points to make sure you cover all of them.

Student answer

The student should have stated that the current used is a.c. (alternating current) as opposed to d.c. (direct current), but is correct to say that it produces a changing magnetic field.

The third statement describes perfectly the purpose of the core.

A transformer consists of a metal core on which are wound two coils of wire, called the primary (or input) coil and the secondary (or output) coil. An electric current applied to the primary coil produces a changing magnetic field in this coil.

The coils are linked magnetically through the core, so the changing magnetic field in the primary coil causes a changing field at the secondary coil. This changing field at the secondary coil causes electromagnetic induction in the secondary coil, which results in the appearance of a voltage across the secondary coil.

This is the basis of the turns-ratio equation:

$$\frac{N_P}{N_S} = \frac{V_P}{V_S}$$

This piece of work would probably score 4 marks out of 6.

The student has correctly described the construction of the coil, but should have indicated that the metal used is soft iron.

The student identifies the process by which a voltage appears across the coil, but has not linked it to number of turns on the coil. The essential feature of a step-up transformer is that $N_S > N_P$.

The student has correctly quoted the turns-ratio equation.

B Peer assessment

2 **Explain how sound waves can be used to detect objects in deep water and measure water depth. In your explanation you must:**

- **refer to the type of sound that is used**
- **refer to the cause of the echo**
- **indicate what must be known about the water**
- **refer to the calculation required.** [6]

Student answer

The sound used to detect objects in deep water is low frequency sound. The sound from the ship passes through the water until it hits a target, such as a shoal of fish. The sound reflects from this target in all directions, but some of it reaches the boat from which it came. This reflected sound is called the echo.

Electronics on board measures the time t between transmission of the sound and receiving the echo. The distance d to the target = speed × time.

Measuring the water depth is an identical process – here the seabed is used as the target.

Use the mark scheme and indicative content to award this answer a level and a mark.

Mark scheme

Level descriptors	Mark
Level 3: Detailed explanation of echo-location. The explanation is substantial and logically structured. At least 6 of the points in the indicative content are covered and the spelling, punctuation and grammar are largely accurate.	5–6
Level 2: The explanation may lack detail, but it is structured and logical. At least 4 of the points in the indicative content are covered and the spelling, punctuation and grammar are usually accurate.	3–4
Level 1: The explanation is quite basic and may be very short and lacks detail and there may be significant irrelevant and incorrect information. At least 2 of the points in the indicative content are covered. There are many inaccuracies in spelling, punctuation and grammar.	1–2
No relevant content	0
Indicative Content: • echo-location uses high frequency sound / ultrasound • sound emitted from ship and strikes a target • reflected sound is an echo • time between sound transmission and return of echo is measured • speed of sound in water must be known • distance $= \frac{1}{2} \times$ speed in water $\times$ time • depth finding uses seabed as the target	

I would give this a level of .... and a mark of ....

This is because ...

....

....

....

C Improve the answer

3 **Use the definition of pressure to explain how the pressure due to a column of liquid in a measuring cylinder depends on the height h of the column.** [6]

Student answer

Pressure is the force acting on a surface divided by the area of the surface.

The column of liquid is a prism of cross section area A and height h.

This means that the weight of liquid is $A \times h$.

So, the pressure depends on the height, because the weight depends on the height.

Rewrite this answer to improve it and obtain the full six marks.

Extended responses: Design, Plan or Outline

The word 'Design' wants you to use your knowledge and experience and be creative in solving an experimental task, while 'Plan' means that you must give detailed information about a procedure or task. 'Outline' means 'summarise', but it is often used to ask students to set out how something should be done (such as 'Outline a plan', 'Outline an experiment', and so on).

> **Tip**
>
> The command words 'design', 'plan' and 'outline' are not exactly interchangeable, but they are often used in a similar fashion in GCSE Physics. Namely, they will be asking you to set out an experiment to test a hypothesis.

A Expert commentary

1 **Design an experiment to show that the extension of a spring is directly proportional to the applied force.** [6]

Student answer

All the major details are listed to enable the experiment to be carried out.

1 Suspend a spiral spring and a metre ruler vertically using a retort stand, boss and clamp.
2 Using the ruler measure the initial length of the spring.
3 Add a 100 gram (1 N) slotted mass and measure the extended length of the spring.
4 Repeat step 3 for loads up to 6 N in steps of 1 N.
5 Record results in a pre-prepared table.

The candidate does not make it clear in that the same initial length is subtracted from the new length to obtain the extension.

6 Calculate the extension for each load, by subtracting the previous extended length from the new extended length
7 Plot a graph of extension against load.

The student omits to say that the line passes through the (0,0) origin. This is essential to establish direct proportion.

8 The line of best fit is strait indicating direct proportion.

There is only one spelling mistake; strait should be straight, but this would not be penalised.

Apparatus

Typical experimental results:

Load (N)	0	1	2	3	4	5	6
Extended length of spring (cm)							
Extension of spring (cm)							

Apparatus:

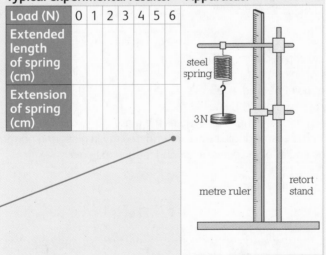

steel spring

3N

metre ruler   retort stand

> **Tip**
>
> A blank table quickly shows an examiner what measurements you intend to make in an experiment.

> **Tip**
>
> Remember a diagram can say a thousand words. Diagrams are often a good way to get information across. But remember to refer to it in your written response to the question.

The student provides a drawing of the apparatus and a table for results, which makes it clear what is described in the written answer. This is good practice.

This piece of work would probably score 5 marks.

B Peer assessment

2 **Outline an experiment to measure the angle of refraction in a rectangular glass block when the angle of incidence in air is 30°. In your answer you must state the apparatus and method you will use. You should also draw a ray diagram to illustrate your plan and indicate the angles of incidence and refraction.** [6]

Student answer

Place a rectangular glass block on a drawing board and draw around its outline with a pencil

Remove the block and draw the normal to one of the long sides

Draw a line, L_1, at 30° to this normal at the point, P, were it meets the glass

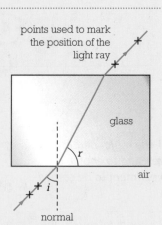

Replace the block and direct a ray of light along the line L_1 and observe the light exit the glass on the opposite side along line L_3

Draw two small crosses on line L_3 and rule a line joining them back to the point Q were the light left the glass

Remove the glass and rule a straight line L_2 between points P and Q

Mesure the angel between line L_2 and the glass. This is the angel of refraction

Use the mark scheme and indicative content to award this answer a level and a mark.

Mark scheme

Level descriptors	Mark
Level 3: Detailed, well-structured plan that would work. At least 6 of the points in the indicative content are covered and the spelling, punctuation and grammar are largely accurate.	5–6
Level 2: The method may lack detail and structure, but with only minor changes it would work. At least 4 of the points in the indicative content are covered and the spelling, punctuation and grammar are usually accurate.	3–4
Level 1: The plan requires significant modification if it is to work. There may be significant irrelevant or incorrect information. At least 2 of the points in the indicative content are covered. There may be inaccuracies in spelling, punctuation and grammar.	1–2
No relevant content	0

Indicative Content:
- set rectangular glass block on paper, outline with pencil
- use protractor to draw normal on one side
- draw line at 30° to normal at point of incidence
- direct ray of light from the ray box along this line
- method to trace emergent ray to point where light leaves the glass
- draw refracted ray in glass and measure angle between refracted ray in glass and normal
- diagram to show normal and angles of incidence in air and refraction in glass

I would give this a level of .... and a mark of ....

This is because ...

....

....

....

C Improve the answer

3 The diagram shows a soft iron plate of weight 5.0 N at rest on a newton balance. Above it is an electromagnet. When the electromagnet is energised, there is an attractive force between the plate and the magnet causing the reading on the balance to change.

Using this apparatus, plan an experiment to show that the strength of an electromagnet is directly proportional to the current in the coil. In your answer you must:

- identify the dependent, independent and controlled variables
- state what measurements you would take
- state what happens to the balance reading when the current is increased
- describe how you would process your results to establish direct proportion. **[6]**

Student answer

1 The dependent variable is the current in the electromagnet. The independent variable is the upward force on the soft iron plate and the controlled variable is the distance between the electromagnet and the soft iron plate.

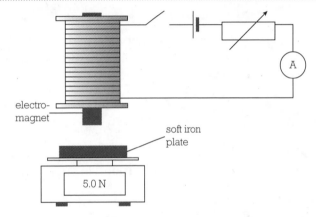

2 Note the balance reading (5.0 N) when the current is zero.

3 Switch on the current and adjust the rheostat until the ammeter reading is 0.5 A.

4 Adjust the rheostat again to increase the current by 0.5 A and record current and balance reading.

5 Repeat step 4 again until a set of balance readings and currents, up to 3.0 A have been recorded.

6 To establish direct proportion, plot a graph of balance reading against current. If the force produced by the magnet is directly proportional to the current, the graph will be a straight line.

Current /A	0.0	0.5	1.0	1.5	2.0	2.5	3.0
Balance reading / N	5.0						

Rewrite this answer to improve it and obtain the full six marks.

Extended responses: Evaluate or Justify

These two command words are slightly different, but both require you to use evidence to make an argument. Both command words may be used together in the same question.

'Evaluate' requires you to use the information supplied in the question, as well as any relevant outside knowledge, to consider evidence for and against an argument. 'Justify', on the other hand, means you need to use the evidence supplied to support and take one argument forward. The main difference between these two words is, therefore, whether you have to assess arguments both for and against a conclusion, or simply the argument for a conclusion.

(A) Expert commentary

1 **A physics student carries out an experiment on refraction using an instrument that can accurately measure angles to 0.1°. The results are shown in the table.**

Angle of incidence i (°)	0.0	10.0	20.0	30.0	40.0	50.0	60.0	70.0	80.0
Angle of refraction r (°)	0.0	6.6	13.2	19.5	25.4	30.7	35.3	38.8	41.0

Evaluate whether or not there is a mathematical relationship between i and r and justify your conclusions. In your response, you should refer to the following:

- **meaning of positive / negative correlation**
- **meaning of direct / inverse proportion**
- **limits of proportionality.** [6]

Student answer

- A positive correlation means that the two variables are equal. Since this is not true here, there is no positive correlation.
- A negative correlation means that one variable is the same as the negative value of the other variable. Since this is not true here, there is no negative correlation.
- Inverse proportion means that when one variable doubles, the other is halved; this clearly does not occur with these data.
- Direct proportion means that the ratio $i:r$ is constant. In the table below this ratio has been calculated to 1 decimal place.

Angle of incidence i (°)	0.0	10.0	20.0	30.0	40.0	50.0	60.0	70.0	80.0
Angle of refraction r (°)	0.0	6.6	13.2	19.5	25.4	30.7	35.3	38.8	41.0
Ratio $i:r$	-	1.5	1.5	1.5	1.6	1.6	1.7	1.8	2.0

The ratio would suggest that for angles of incidence less than 30°, there is direct proportion between i and r, but only up to a limit of proportionality at about $i = 30°$

This work would probably score 4 marks.

The student does not understand positive and negative correlation, and would not get any credit.

The student has a clear understanding of direct and inverse proportion and would gain full credit.

The student suspects direct proportion, has calculated the correct ratio to test for it, and entered the values in the table.

The student has realised that there is now evidence of limited direct proportionality, and the calculations have provided the justification.

The student has reached the correct conclusion and provided adequate justification.

B Peer assessment

2 The volume of materials change with temperature. The graph
   shows how the volume of a fixed mass of liquid water changes as its
   temperature rises from 0 °C to 4 °C and then to 100 °C.

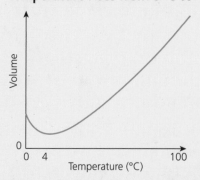

> **Tip**
> Always look carefully
> at graphs and identify
> their main features
> before starting your
> answer.

Study this graph showing changes in the volume of the liquid water
over a given temperature range. Determine what is happening to the
water's physical properties. Justify your reasoning.

In your answer you must refer to what is happening to the changes in:

- the separation of the molecules
- the density of the water
- the forces of attraction between water molecules.   [6]

Student answer

1 As the temperature rises from 0 °C to 4 °C, the volume falls. This implies
   that the average separation of the water molecules is decreasing.

2 As the temperature rises from 4 °C to 100 °C the volume rises. This
   implies that the average separation of the water molecules is increasing.

3 At 100 °C the water changes into steam. Gases have lower densities than
   liquids, so the density must be decreasing.

4 At 0 °C the water changes into solid ice. Solids have higher densities
   than liquids, so the density must be increasing.

5 Between 0 °C and 4 °C the volume is falling, so the attractive forces
   pulling the molecules together are increasing.

6 Between 4 °C and 100 °C the volume is rising, so the attractive forces
   pulling the molecules together are decreasing.

Use the mark scheme and indicative content to award this answer a level and a mark.

Mark scheme

Level descriptors	Mark
Level 3: Detailed, well-structured argument given. At least 6 of the points in the indicative content are covered and the spelling, punctuation and grammar are largely accurate	5–6
Level 2: The argument may lack detail and structure. At least 4 of the points in the indicative content are covered and the spelling, punctuation and grammar are usually accurate.	3–4
Level 1: The argument is broadly accurate. There may be significant irrelevant or incorrect information. At least 2 of the points in the indicative content are covered. There may be inaccuracies in spelling, punctuation and grammar.	1–2
No relevant content	0

Indicative content:

The fall in volume from 0 °C to 4 °C implies:
- a decrease in the separation of the molecules
- an increase in the density, because the mass is constant
- an increase in the attractive forces between molecules

The rise in volume from 4 °C to 100 °C implies:
- an increase in the separation of the molecules
- a decrease in the density, because the mass is constant
- a decrease in the attractive forces between molecules

The water has:
- maximum density at 4 °C

I would give this a level of .... and a mark of ....

This is because ...

....

....

....

C Improve the answer

3 **A radioactive source is placed in front of three absorbers. A detector of radiation is placed at points A, B, C and D. The detector at B detects radiation after it passes through the paper and aluminium, but not lead.**

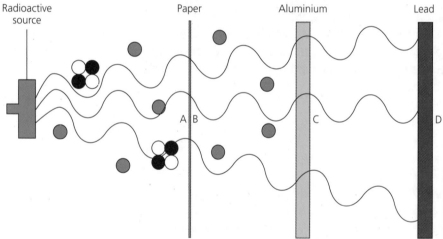

Radioactive source   Paper   Aluminium   Lead

The count rates, after correcting for background radiation, are recorded at these four locations and are shown in the table.

Detector placed at	A	B	C	D
Count rate in cpm	255	154	149	74

Identify what type(s) of radiation are emitted from the source. Justify your answer. [6]

Student answer

Lead absorbs gamma. The results therefore point to the source being an emitter of gamma radiation.

Rewrite this answer to improve it and obtain the full six marks.

Tip
Remember that 'justify' might appear with other command words such as 'determine' and 'identify'. Make sure you read every question thoroughly so you know exactly what you're being asked.

Tip
When asked questions requiring you to justify your answer, quote the relevant information you are given in the question that leads you to your conclusion.

3 Working scientifically

Working scientifically is about what scientists do and how they should work in a scientific investigation. It is a required part of GCSE Physics, although you will never be asked questions that are specifically flagged 'working scientifically' in an exam.

Being able to work scientifically is a skill, not an assessment objective. It is about thinking like a scientist. This mindset can be hard to get into, but once you start thinking like a scientist, it will be a useful skill for GCSE Physics and beyond.

Working scientifically skills should be covered as you work naturally through your course. This section should make you aware of these skills, so you can look out for where they crop up in your studies, and use these opportunities to develop your scientific brain.

Working scientifically includes several different skills, which fall broadly into the following areas.

1 The development of scientific thinking

2 Experimental skills and strategies

3 Analysis and evaluation

4 Vocabulary, units, symbols and nomenclature

This section will just deal with the first three areas, as the fourth area – vocabulary, units, symbols and nomenclature – is covered in the Maths and Literacy sections of this book.

» Apparatus and techniques

As well as the four working scientifically areas, you are required to demonstrate your ability to use a range of apparatus and techniques. These are also known as AT skills.

Table 3.1 GCSE Physics AT skills

AT 1 Use of appropriate apparatus to make and record a range of measurements accurately, including length, area, mass, time, volume and temperature. Use of such measurements to determine densities of solid and liquid objects
AT 2 Use of appropriate apparatus to measure and observe the effects of forces including the extension of springs
AT 3 Use of appropriate apparatus and techniques for measuring motion, including determination of speed and rate of change of speed (acceleration/deceleration)
AT 4 Making observations of waves in fluids and solids to identify the suitability of apparatus to measure speed/frequency/wavelength. Making observations of the effects of the interaction of electromagnetic waves with matter
AT 5 Safe use of appropriate apparatus in a range of contexts to measure energy changes/transfers and associated values (e.g. work done)
AT 6 Use of appropriate apparatus to measure current, potential difference (voltage) and resistance, and to explore the characteristics of a variety of circuit elements
AT 7 Use of circuit diagrams to construct and check series and parallel circuits including a variety of common circuit elements
AT 8 (single sciences only) Making observations of waves in fluids and solids to identify the suitability of apparatus to measure the effects of the interaction of waves with matter

As with the rest of the working scientifically criteria, you will likely cover these AT skills as part of your course. However, it helps if you're aware of them as you work through your specification. Examples of how the AT skills might tie in to your learning are given in Table 3.2.

Table 3.2 How AT skills tie into your learning

	Measurement	What you might use (apparatus)	What you might do (techniques)
AT 1	length, area, mass, time, volume and temperature, density	ruler, balance, measuring cylinder, burette, thermometer	finding the density of regular and irregular solids and liquids
AT 2	forces	springs, newtonmeters, slotted masses on carriers	Hooke's Law experiments, construction of a newtonmeter
AT 3	motion, including determination of speed and rate of change of speed (acceleration /deceleration)	stopwatches, metre rules, dataloggers (with computers), linear air tracks	finding the speed and acceleration of trolleys on a runway or in collision
AT 4	speed, frequency, wavelength of waves and observation of their interaction with matter	ripple tanks, stroboscopes, stretched strings, CRO, black-bulb thermometers, microphones, loudspeakers	finding the properties of water waves and sound waves
AT 5	energy changes, transfers and associated values such as work done	joulemeters, ammeters and voltmeters	finding specific heat capacity, electrical power of a motor and the efficiency of a machine
AT 6	current, potential difference (voltage) and resistance	ammeters, voltmeters, multimeters, resistance wire, thermistors, filament lamps, diodes	finding $V-I$ characteristic graphs of various discrete components
AT 7	voltage and current	voltmeters, multimeters	finding the current flowing in series and parallel circuits and the voltage at key points
AT 8	waves	water in *Perspex* containers, rectangular glass blocks, ray-boxes, microphones, loud speakers, metal rods, CROs, silvered and black painted metals and radiant heat sources	studying the passage of light through water and glass and measuring the speed of sound in air and a metal rod using a CRO; studying the absorption and radiative powers of different types of surface

Most of these AT skills will be demonstrated as part of your required (or core) practicals, but you can look out for opportunities to demonstrate these skills in any experiments you complete.

» The development of scientific thinking

This area of working scientifically is about understanding how scientific theories come to be developed and refined as well as recognising the importance of working safely and the limitations of what we can discover.

How scientific methods and theories develop over time

You may have heard the term 'scientific method' before – the scientific method is essentially a recipe for problem solving. It starts with an observation or a question such as 'what makes ships float?'. This question then leads to ideas and suggestions (scientists call these hypotheses), and possibly predictions. This in turn leads to experiments to test the hypotheses and predictions. If the results are consistent with the predictions, we have the start of a theory. Otherwise, we must review the hypothesis and start over again.

> **Key term**
>
> Hypothesis: A suggestion made, with limited evidence, to explain a phenomenon or observation that can then be tested through practical means. The plural of hypothesis is hypotheses.

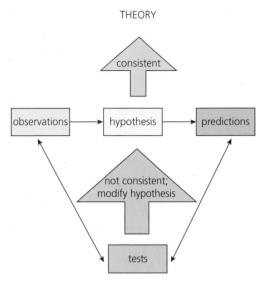

▲ Figure 3.1 The scientific method

A scientific theory is a hypothesis (or a group of hypotheses) about some phenomena that are supported through experimental research. You will know about some of these already, such as atomic theory and kinetic theory. Theories are important because they explain physical phenomena, and are subject to test. If there is no way to test an idea, it is not a theory. If it can be tested, and passes the test, it becomes a good theory. You need to know about theories because they show how understanding of the physical world has developed and how they inform the way scientists currently think.

Over time good theories become scientific laws. For example, Newton first had a theory about motion. Over time his ideas were summed up in 'Newton's Laws of Motion', with which you are probably familiar.

Figure 3.2 shows an example of how one theory (specifically atomic theory) developed over time.

> **Key term**
>
> Phenomenon: An observation that prompts you to ask questions. The plural of phenomenon is phenomena.

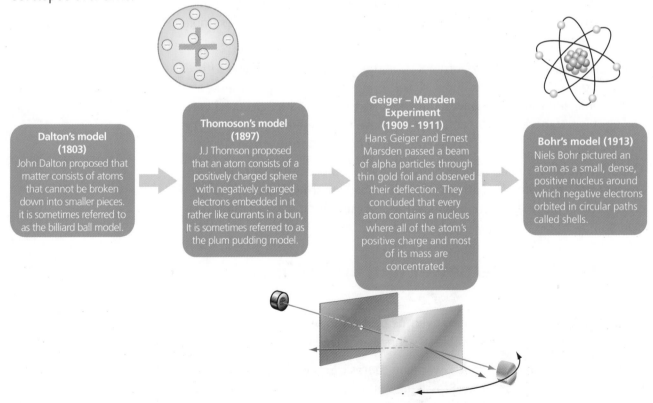

▲ Figure 3.2 Development of atomic theory

Science develops when challenging questions are asked. For example, you might ask why the positive protons inside a nucleus don't repel each other. It was only half a century after Rutherford's model was proposed that Gell-Mann found an answer to that question thanks to the development of particle accelerators.

Science also develops when there is an improvement in technology. We know that, normally, gases do not conduct electricity. But, in 1911, Kamerlingh Onnes discovered that when helium was cooled to a temperature of about −269 °C, it not only conducted electricity, but it had no electrical resistance whatsoever. Onnes' discovery was only possible because of the development of techniques to cool gases to these very low temperatures. Onnes' discovery in turn led to the development of modern MRI scanners used in hospitals throughout the world.

Tip

You may get a question asking you to evaluate a hypothesis or theory. If so, you will need to apply your knowledge and understanding of natural phenomena to support or undermine the theory.

Using models to solve problems, make predictions and develop explanations

There are several types of model used to make predictions and develop explanations. The models you need to know are:

- Descriptive
- Mathematical
- Representational
- Spatial

A **descriptive model** is like a picture you hold in your mind about the physical world. Examples include the particle model of atoms in liquids, or the model of alpha particles knocking out electrons as they ionise air. You needed a descriptive model on page 74 when answering a question on liquid volume.

A **mathematical model** describes physical phenomena as a series of equations. You will frequently apply mathematical models in physics. For example, all the equations you use to solve questions about electricity are based on the mathematical model of moving electrons. Newton's Laws are another example of a mathematical model.

A **representational model** represents something that we cannot see, with something we can see. For example, slinky springs are often used to show longitudinal and transverse waves. You know that sound is a longitudinal wave and light is a transverse wave although you have never *seen* the vibrations that cause them. Instead, the motion is represented by the motion in the slinky.

A **spatial model** is similar to both descriptive and representational models. For example, the Rutherford atom can be described in terms similar to our description of planetary motion and we can represent it on a two-dimensional diagram.

Appreciating the power and limitations of science

Science has solved many problems over the past centuries, and will solve many more in the future. But some problems cannot be solved by science alone.

We know, for example, that global warming causes climate change, and that human activity – such as burning fossil fuels and cutting down rainforests – increases global warming. However, science cannot prevent people from burning coal to keep warm in the winter, or stop all buses, trains, cars and planes. That is a decision for people, countries, governments and the United Nations. Science can provide alternatives – such as alternative energy sources that are cleaner and safer – but these alternatives still have to be embraced by the world. There is a limit on science's reach and power.

There are also limitations on scientists. Many scientists recommend increased use of nuclear power. But events in Chernobyl and Fukushima have made governments reluctant to use *any* nuclear energy.

Finally, there are ethical issues that scientists must consider. For example, scientists may question whether they should manufacture weapons or develop methods to defend against nuclear attack.

> **Key term**
>
> Ethical issues: Issues that involve deciding whether a course of action is morally right or wrong.

Other contentious ethical issues include genetically modified food, cloning and 'designer babies'. Many of these areas are governed by law, which provides guidance on what is and isn't ethical (although some people argue that these laws limit what scientists are able to do).

For new scientific developments in areas that outpace the speed of law, scientists have to take the difficult ethical decisions themselves.

Evaluating risks in science

Whenever scientists do practical work they need to evaluate risks – these risks could be minor and limited, or potentially catastrophic.

You will be familiar with the need to evaluate risk and should have carried out risk assessments when doing practical work. Risk assessments are an attempt to safeguard yourself, your classmates and the apparatus. All experiments carry risk – even if it is only the risk that someone will, for example, swallow something.

A risk assessment requires you to:

1 think about **what** might reasonably go wrong

2 decide **how** likely it is to go wrong

3 consider **the effects** on people and equipment if the problem occurs.

Here is how to complete a risk assessment.

1 Consider the physical hazards in the laboratory. For example, the things that you can trip over, electrical leads trailing from one bench to another, and so on. Make sure you attend to them before doing any practical activity.

2 Identify specific hazards that relate to the experiment. These might be electrical (burns and shocks), radioactivity (exposure to radiation), high or low temperatures (burns and scalds), light (damage to eyes from lasers) and so on. These experiment-specific hazards are those that you need to identify in a GCSE exam.

3 Think about the effects of these hazards. They should be listed in order of importance.

- Death or permanent disability
- Long-term illness or serious injury
- Medical attention required
- First aid required

It is important that you are aware of all the possible risks and take action to avoid them. The greater the hazard, the more important it is that you take action to eliminate or reduce the risk.

A few examples of hazards, risks and precautions are shown in Table 3.3.

Table 3.3 Examples of hazards, risks and precautions

Hazard	Risk	Safety precaution
long hair	could catch fire	tie back long hair
hot apparatus	could cause burns	allow to cool before touching; use tongs
flying springs	could cause blindness	wear safety goggles
radiation	over-exposure	use computer simulation instead

Recognising the importance of peer review

Before publication of experimental results and new theories, scientists always present their ideas at conferences and have their work checked and evaluated by other experts in the same field. This is called peer review.

In 1989, Martin Fleischmann and Stanley Pons reported that they had carried out electrolysis experiments, which produced so much excess heat that the results could only be explained in terms of nuclear fusion. This, they claimed, was fusion at room temperature.

By 1991, more than 600 scientists were working on 'cold fusion', trying to reproduce the results of Fleischmann and Pons. Today, it is commonly thought that the excess heat was not caused by nuclear fusion, but by contaminants in the electrodes.

It is not enough to just announce what you have done in science – others must be able to reproduce your results as well.

Key term

Peer review: The process in which experts in the same area of study read, consider and report on the findings of another scientist before it is considered for inclusion in a scientific journal (magazine read by scientists).

Questions

1 What is a scientific hypothesis?
2 How do scientists usually test a hypothesis?
3 What do physicists do when one experiment after another suggests their theory is wrong?
4 A student wrongly claims that working with a 9V battery, a lamp and some connecting wire is totally without risk.
   Why is the student wrong?
5 What is the main hazard when doing an experiment involving the stretching of springs and how can the risk be reduced?
6 Many schools have eliminated the risks associated with having pupils carry out some quite hazardous experiments by showing experimental demonstrations.
   What else might they do?
7 A student is about to carry out an experiment to find the energy needed to change liquid water into steam.
   What would be the major hazard in this experiment?
8 There is a legal requirement that all mains powered electrical equipment in schools is checked annually.
   Suggest why this is so.
9 A teacher insists that all schoolbags are stored under the benches before any practical work is carried out.
   a Why are schoolbags a hazard?
   b Why is the teacher's action good practice?
10 What is peer review and who carries it out?
11 Some people say that drinking any amount of alcohol and driving should be an offence. Others argue that it should not.
   Explain whether or not this is a problem science can solve.

» Experimental skills and strategies

Being a scientist and carrying out practical experiments means that you need to use appropriate experimental skills and strategies to ensure that your results are meaningful. This means developing a hypothesis, then working out an experiment on how best to test this hypothesis. You may be familiar with the sorts of questions you will be asked in the exam around planning or outlining an experiment. In this section, you will be shown best practice for any experiment.

Developing hypotheses

Science is all about observations and asking questions. For example, imagine you notice that a pendulum clock is losing time. You might first ask 'Why is this happening?'. You guess that the time for the pendulum to make an oscillation (its period) depends on the weight at the end of the pendulum. This is your first hypothesis. To test this hypothesis, you carry out an experiment.

It turns out that your first idea is wrong. So, you put forward another idea – that the period depends on the pendulum's length. This is your second hypothesis. You carry out another experiment and find that you are right.

The important thing about a good hypothesis is that an experiment can be designed to test it – not whether it is correct.

Planning experiments to test hypotheses

As discussed, we test a hypothesis by planning and carrying out an experiment. The first thing to ask is 'What do we want to find out?'.

In the pendulum example on the previous page, the second hypothesis suggested that changing the length of the pendulum would change its period. We call the period the dependent variable, because it depends on something. We *think* it depends on the length of the pendulum – so the length of the pendulum is the independent variable. This is because this variable is independent of the experiment – it is changed by the scientist.

It is important that scientists are specific about what they are testing. Changing more than one thing each time might give misleading results. For example, if we think that the period might also depend on the mass of the pendulum, the mass must not change (it must remain constant) when we are testing the length. The mass is, therefore, a controlled variable. Keeping the controlled variables constant makes it a fair test.

In any science experiment there should only be **one** dependent variable and **one** independent variable. All the other variables must be controlled.

There are two other types of variable you need to know about. A continuous variable has values that are numbers. Mass, temperature and volume are examples of continuous variables. The variables used in physics experiments are almost always continuous variables.

A categoric variable is one that is best described by words. Variables such colour, shape and type of car are categoric.

> ### Key terms
>
> Dependent variable: The variable that changes because of the change in the independent variable.
>
> Independent variable: The variable that the physicist decides to change.
>
> Controlled variables: The variables that are kept constant throughout an experiment.
>
> Fair test: A test in which there is one independent variable, one dependent variable and all other variables are controlled.
>
> Continuous variables: The variables that can have any numerical value (such as mass, length).
>
> Categoric variables: Variables that are not numeric (such as colour, shape).

> ### Tip
>
> For more information on the types of apparatus and techniques you'll need to know see page 74.

Choosing appropriate apparatus and techniques

Whatever experiment you are carrying out, the results will only be useful if you have selected the appropriate tools.

For example, when testing the length of a pendulum and its period, you will need to measure length, time and weight. So, you need to choose the most appropriate equipment to measure these quantities, as well as decide how best to use the equipment.

For example, a metre stick is appropriate to measure the length as it is unlikely to be longer than this and you can see the length to approximately 1 mm. To use this metre stick correctly and ensure the test is fair, you have to measure the length in the same way each time. To do this you should ensure the object you are measuring the length of is placed exactly alongside the metre stick and that both the object and metre stick are straight.

You need to consider whether there are any other obstacles to a fair measurement. For example, making sure there are no knots in the pendulum.

You must also make decisions about different techniques, such as what is the best length to measure. You could measure from the point of suspension to the bottom of the object, or from the point of suspension to the middle of the object. Again, you need to think scientifically. Weight acts from the centre of gravity, which is in the middle, so the best technique is to measure to the middle of the object.

Measuring time is less accurate because there will always be an element of reaction time. To ensure that this effect is minimised, you need to think scientifically once more. You could pick a more accurate stopwatch (one capable of measuring to at least 0.1 s is probably suitable), or you could time the period after allowing the pendulum to make a few swings first – starting the stopwatch when it reaches the end of a swing. It is also good practice to repeat the timing a few times and find the average period.

Table 3.4 lists a few common pieces of measuring apparatus found in a physics laboratory and what they are used to measure. Part of your training is to develop the technique to use them properly.

Table 3.4 Common apparatus in a physics laboratory

Instrument	measuring cylinder	newtonmeter	protractor	thermometer	ammeter	voltmeter
Measures	liquid volume	force	angles	temperature	current	voltage

Carrying out experiments accurately

The accuracy of any experiment you do depends on two factors: the equipment used and your technique. When thinking about how to carry out an experiment accurately, you need to know the difference between accuracy, reliability, precision and resolution.

Accuracy is how close we get to the true value of a physical measurement. For example, suppose five students measure the length of the same metal rod. They all use a metre rule with a centimetre scale. They get the following results:

91 cm, 93 cm, 90 cm, 92 cm, 89 cm

> **Tip**
>
> You should be able to identify the independent, dependent and controlled variables in every investigation you carry out. This is important because variable identification is a question that frequently appears in GCSE exam papers.

> **Key terms**
>
> Accuracy: Accuracy is how close we get to the true value of any physical measurement.
>
> Reliability: A test is defined as reliable if different scientists repeating the same experiment or measurement consistently get the same results.
>
> Precision: Precision measures the extent to which measurements are the same.
>
> Resolution: Resolution is the fineness to which an instrument can be read.

The **range** of these results is calculated as:

highest value – lowest value = 93 – 99 = 4 cm

The **mean** of these results is 91 cm.

It is likely that some of the students obtained a figure higher than the true value, while some obtained a figure that was lower. By taking the mean we effectively cancel out the too-high values with the too-low values. Our best (most accurate) value for the length of the rod is, therefore, 91 cm. Remember that to improve accuracy you should **repeat** and then **average**.

Another way to improve accuracy is to **use a better measuring instrument**. A digital voltmeter, for example, is likely to be more accurate than an analogue meter.

Resolution is the fineness to which an instrument can be read. For example, a stopwatch with a sweeping hand might have a resolution of $\frac{1}{10}$ of a second, while a digital stopwatch might have a resolution of $\frac{1}{100}$ of a second. However, both stopwatches have the same precision because this factor will be determined by the reaction time of the person using it.

Precise measurements are those where the range is small. For example, suppose three students measure the mass of a beaker using balance A. They get the results 45 g, 39 g and 42 g. The range of these measurements is 45 – 39 = 6 g, and the mean is 42 g.

Suppose they repeat the measurements using the same beaker, but a different balance, balance B. They get the results 42 g, 41 g and 43 g. The range of these measurements is 43 – 41 = 2 g, and the mean is 42 g. The readings on both balances have the same accuracy, but those using balance B have greater precision.

A test is defined as **reliable** if different scientists repeating the same experiment or measurement consistently get the same results. The technique **to improve reliability is to repeat the same test several times**.

accurate
and precise

precise, but
not accurate

not accurate,
not precise

▲ Figure 3.3 Accuracy and precision

Making and recording observations

You need to know how to record the results of an investigation. In almost all practical work in physics you record results in a table.

When drawing tables and recording data ensure that:

- the lines in your table are drawn with a ruler and pencil
- there are headings for each column and/or row
- there are units for each column and/or row –usually placed after a solidus or within brackets after the heading; for example 'mass / g' or ' current (A)'
- units are not written beside the numbers in the table
- there is enough space for repeat measurements and averages – remember the more repeats you do, the more reliable the data
- data items are recorded to the same number of decimal places or significant figures.

For examples on how this is done in practice refer to the Maths skills section of this book.

Evaluating methods and suggesting possible improvements

Evaluating means assessing how it is going as you carry it out, and, at the end, thinking about what could be improved if it was to be carried out again. You should always be evaluating while carrying out practical work.

Evaluating is an important part of the scientific method. Written evaluation is often required as a part of a conclusion.

During an experiment, you may find that some ideas, apparatus or techniques are not working well. If you need to make changes to improve your method, do so and refer to it in your evaluation. Scientists have to be flexible – but you need to explain *why* you changed your plan. Remember that you will need to repeat all your tests whenever you change your approach to ensure it is fair.

As part of evaluating, you need to ask questions such as:

- was the experimental method suitable
- could it have been better and, if so, how
- were there any sources of error
- how could the method be improved to reduce, or eliminate, known sources of error?

> **Tip**
> Evaluation is a skill that you should apply to every aspect of your work, but particularly in required practicals and exam questions that ask you to 'evaluate'.

For example, in experiments to measure specific heat capacity, heat energy is generally lost to the environment, even with good insulation. To reduce the impact of this heat loss, it is usually better that the material being heated is first cooled to a few degrees below room temperature and then heated to a few degrees above room temperature. That way, the material being tested is likely to gain as much heat from the environment in the first part of the procedure as it loses in the second part.

Questions

1 Identify the independent variable, dependent variable and **one** controlled variable for each of the following investigations.
   a An experiment is carried out to see if there is any link between the resistance of a piece of wire and its length.
   b A student is investigating Newton's Second Law. The student wants to find out the relationship between the applied force on a trolley and its acceleration.
   c In an experiment to measure the specific heat capacity of aluminium, electrical energy is supplied and the rise in temperature is noted.
   d A student investigates whether there is a link between the time it takes a marble to roll down a slope and the vertical height of the ramp.
   e A mass is attached to the end of a spiral spring, which is suspended vertically. A student investigates the relationship between the periodic time of the oscillations and the mass attached to the spring.
   f A student knows that when an electric current flows in a metal wire heat is produced. The student investigates the relationship between the amount of heat produced and the current in the wire and the amount of heat produced per second.

2   A student balances a metal rod at its midpoint and the rod comes to rest in a horizontal position. The student then attaches a weight W at a point 25 cm from the midpoint. The student investigates at what distance d from the midpoint a load of 5 N must be placed on the other side of the midpoint to restore equilibrium, for various loads W ranging from 1 N to 20 N.

   a Using the information provided, write a suitable hypothesis for this investigation.

   b State how the student will recognise when equilibrium has been restored.

   c The student finds that when W is greater than 10 N the metal rod cannot be restored to equilibrium. Does this observation make your hypothesis invalid?

3   State the most suitable apparatus for measuring each of the following quantities.

   a mass of about 25 grams to a precision of 0.1 gram

   b volume of 24.7 cm^3 of liquid

   c time of about 45 s to the nearest 0.1 s

   d current of about 0.050 A to the nearest 0.001 A

   e temperature of about 80 °C to the nearest degree

4   A student measures the acceleration due to gravity and obtains the following five results: 9.7 m/s^2, 9.8 m/s^2, 9.8 m/s^2, 9.8 m/s^2, 10.1 m/s^2

   a Calculate the range and mean of these results.

   b Why is the mean considered to be the most accurate measurement?

5   State the main advantage of researchers in physics repeating their measurements.

6   A physics student measures out five different volumes of ethanol and measures the mass of each. The results obtained are:

   20 cm^3 16 g; 35 cm^3 28 g; 45 cm^3 36 g; 50 cm^3 40 g; 55 cm^3 42 g

   a Present these results in a suitable table with headings.

   b Which result might the student want to repeat? Why?

7   Three students are designing an experiment to measure the time taken for a trolley to roll down a ramp. They consider three timing methods:

- using a stopclock capable of measuring time to the nearest second
- using a stopwatch capable of measuring time to $\frac{1}{100}$ th of a second
- using a light gate and datalogger, capable to measuring time to $\frac{1}{100}$ th of a second

The true time to run down the ramp is 9.5 s.

   a Which instrument has the least resolution?

   b The students all decide to use the datalogger method and, independently, they obtain the following times (to 1 dp):

Student	Time (seconds)				
A	9.8	9.3	9.9	10.3	10.3
B	9.8	9.8	9.8	9.9	9.8
C	9.5	9.4	9.6	9.5	9.5

   i Which student's results have the greatest precision?

   ii Which student's results are neither reliable nor precise?

   c Student A repeats the experiment. Why is this a good idea?

» Analysis and evaluation

Collecting, presenting and analysing data

When carrying out physics experiments you will often record your results in a table. However, it can be difficult to spot trends from a table, so you may decide to create a graph. Once in graph form, the gradient of the line and its intercept on the vertical axis can provide a more obvious indication of a trend. For more details on graphs, tables, distributions and analysing data, see the Maths skills section of this book (page 5).

Evaluating data

Accounting for uncertainty

All data collected by a scientist contains an element of uncertainty. This can be due to the instrument's lack of precision or inconsistencies in measurements made by the individual.

Uncertainty in a measurement is the maximum difference between the mean value and the experimental values.

If a student makes three measurements of the density of a liquid and obtains values (in g/cm^3) of 1.48, 1.53 and 1.49, the mean is $1.50\,g/cm^3$. The uncertainty is calculated as $1.53 - 1.50 = 0.03\,g/cm^3$. A large uncertainty means poor precision.

This is a problem because we cannot tell which of the values taken is best for our measurement.

The cause of experimental uncertainty is known as error. There are two types of error – random error and systematic error. Make sure you are aware of them so you can take steps to eliminate or reduce their effects.

Random error

A random error is one that causes a measurement to differ from the true value by different amounts each time. Three students measuring the volume of a cube are likely to come up with three different values. This is the result of random error. The error is randomly scattered about the true value. By making more measurements and calculating a new mean, we reduce the effects of random error.

Systematic error

A systematic error is one that causes a measurement to differ from the true value by the same amount each time. This will usually be due to the equipment used. If you have ever plotted a graph and found that it crossed the vertical axis when you expected it to go through the origin, the cause is likely to be a systematic error as *all* the results are different by the same amount. Systematic errors cannot be dealt with by simple repeats. Instead, you need to use a different technique or apparatus.

> **Key terms**
>
> Random error: An error that causes a measurement to differ from the true value by different amounts each time.
>
> Systematic error: An error that causes a measurement to differ from the true value by the same amount each time.

> **Tip**
>
> A good straight line graph that does not pass through the origin, when it is expected to do so, is nearly always the result of a systematic error.

The following data illustrates random error in an experiment measuring the time to fall to earth from different heights. To help you recognise the random nature of the error, the expected time is given for comparison. The measured time is sometimes bigger, sometimes smaller than expected. This is characteristic of random error.

Height (m)	1.0	1.5	2.0	2.5	3.0
Measured time to fall (m)	430	571	622	731	752
Expected time (s)	*452*	*553*	*639*	*714*	*782*

The data and graph below illustrate systematic error in an experiment on Hooke's Law with a spring.

Force (N)	2	4	6	8	10
Extension (mm)	50	90	130	170	210

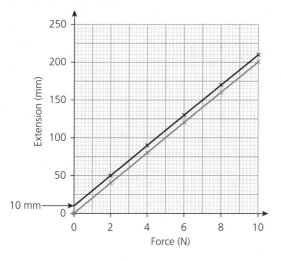

The extension:force ratio is not constant as expected. A graph of *extension* against *force* plotted with these figures is a straight line, but it cuts the vertical axis at 10 mm.

It appears that all the extension values are 10 mm larger than they should be. The experiment should, therefore, be repeated, this time adding a load at the start to remove imperfections in the spring (or perhaps changing the spring for a new one).

Checking for repeatability, reproducibility, validity and reliability

Results should be repeatable, reproducible and valid to ensure they are reliable when evaluating a hypothesis or developing a conclusion.

Reliable results are important in assessing whether we have discovered something meaningful. If results were, for example, repeatable but not reproducible, or reproducible but not valid, the results may be incorrect. Results that are simply repeatable are particularly untrustworthy as the person carrying out the experiments may just be repeating their mistakes again and again. Reproducible results give greater confidence that they are correct because several people or techniques are involved, but if the wrong thing is being investigated it is still not valid or meaningful.

Key terms

Repeatable: Results are repeatable if similar results are obtained when an experiment is repeated by the same person several times.

Reproducible: Results are reproducible if similar results are obtained when the experiment is repeated by another person or by using a different technique.

Valid: Results are valid when the measurements are correct measures of the property being investigated. For example, measuring the length of a magnet is not a valid way to measure its strength.

Reliable: Results are reliable if they are valid, repeatable and reproducible.

Tip

It is probably random error if some of the points on your graph are above the line of best fit while others are below it.

Questions

1 Two students measure the resistance of the same length of wire four times and obtain the values shown in the table. The true resistance is 5.6 Ω.

Student	Resistance (Ω)			
A	5.5	5.7	5.8	5.4
B	5.7	5.7	5.7	5.8

  a For each student, calculate the uncertainty in the mean.
  b Compare the accuracy of each student's results.

2 Decide whether each of the following will lead to a systematic or random error.
  a Using an analogue ammeter with a zero error.
  b Forgetting to re-zero the balance in one repeat experiment.
  c Finding the density of saltwater when asked to find the density of ethanol.

3 Explain the difference between an *error* and a *mistake* when doing an experiment.

4 Two students present the results of their experiment in the form of a graph. The correct graph is a straight line through the origin with a gradient of six units.
  Student A gets a straight line of gradient six units but it does not pass through the origin.
  Student B gets a straight line passing through the origin, but its gradient is eight units.
  What type of error has probably been made by:
  a Student A?   b Student B?

5 What type of error does the technique of *repeat and average* reduce and why?

6 A wire is 98.4 cm long. A student measures the length with four different metre sticks.
  The results are: 96.8 cm, 97.1 cm, 97.7 cm, 97.9 cm
  a What type of error has been made by the student?
      Give a reason for your answer.
  b Three other students measure the same length of wire with different metre sticks and they all get 98.4 cm.
      Which two of the following words apply to these measurements?
      reproducible, repeatable, valid

≫ Scientific vocabulary, quantities, units, and symbols

Science vocabulary

Science has its own vocabulary. Words used often have a very specific meaning. This is why it is important to learn the definitions of technical terms to understand and create your own written science. A common mistake is to use words as they are understood by the non-scientist. For example, stating that the weight of an object is 50 kg, when you should be talking about mass. For information on scientific vocabulary used in physics, refer to the Literacy section of this book.

Science quantities, units and symbols

For information on the quantities and units used in science, as well as how to convert between them, refer to the Maths section of this book.

4 Revision skills

This section covers the importance of revision and the key strategies that can help you gain the most benefit from your revision. A common misconception is that there is only one way to revise – one that involves lots of note-taking, re-reading and highlighting. However, research shows that this is not an effective way of revising. You need to vary the techniques you use – and find the ones that work best for *you* – to make the most of revision.

Students often think they can change the way they revise, or that revision is something you either can or cannot do. In fact, revision is an important skill and, like any skill, with support and practice you can get better at it. By 'better', this means that you can revise more efficiently (in other words, you'll get a greater benefit from the same amount of revision time) and more effectively (in other words, you'll retain more information).

This chapter will cover the key elements of successful revision:

- Planning ahead
- Using the right tools
- Creating the right environment
- Useful revision techniques
- Practice, practice, practice!

>> Planning ahead

The key to successful revision is planning. There are a number of things to bear in mind when planning revision.

Be realistic

There is nothing more demotivating than setting unrealistic targets and then not fulfilling them. You need to think carefully about how much work you can realistically complete and set a reasonable time to complete it.

Ensure you cover all topics in the course

It is tempting to focus on what you think are the most important areas and leave out others. This is risky because no one knows what you will be asked about. It's a horrible feeling seeing an exam question on a topic that you know you haven't revised. In this section, there is advice on the sorts of strategies to ensure you cover all the key points of the specification.

Make friends with the areas you don't like

It is tempting to focus on the areas you already know and are good at. It makes you feel like you're making great progress when, in fact, you're doing yourself a disservice. You should work hard at the areas you find difficult to make sure you give yourself the best chance. This can be tough as you may feel progress is slow, but you must persevere with it.

> **Tip**
> Spend a small amount of time each evening during your GCSE course going over what you learnt in that day's lesson – it can be really beneficial. It helps you remember the content when you come to revise it, and provides good preparation for the next lesson.

Ask for help

The most successful students are often those who ask questions from teachers, parents and other students. If there is anything on the specification that you are unsure about, don't stay silent – ask a question! Proper planning will ensure you have time to ask these questions as you work through your revision.

Target setting

Targets are an important part of successful revision planning. You may want to include SMART targets in your revision timetable.

Here's an example of a SMART (specific, measurable, achievable, realistic and timely) target.

Target: Achieve at least a grade 6 in a practice Chemistry Paper 1 done under exam conditions. This should be completed by the end of the week.

- **Specific** – this target is specific as it gives the exam paper, how it needs to be completed and the grade required.
- **Measurable** – as a specific minimum grade is given (6), this target is measurable.
- **Achievable** – as long as there is time to complete the paper, which there should be if it's being completed in the 'time allowed', then this target would be achievable.
- **Realistic** – you shouldn't be expecting to score grade 9 in assessments straight away or learn huge amounts of content in a very small time; so, a grade 6 seems to be realistic for a first stab.
- **Timely** – there is a set time to complete this goal, namely by the end of the week. Assuming that the student has revised all of the topics on this paper by then, this is a sensible timeframe.

Targets can also be smaller and set for individual revision sessions, for example:

- complete three practice questions on one maths skill
- get 75% on a recall test
- learn the stages of a process, e.g. the carbon cycle
- make a set of key word flash cards on Lenses and Visible Light

Setting targets for each revision session will help you realise when you are finished, as well as providing yourself with evidence of your progress – always a good motivator!

> **Tip**
> Targets can include things such as not using social media or your phone for a whole revision session if this is something you particularly struggle with.

❯❯ Using the right tools

Having the right tools is vital for effective revision. Some of the 'practical' tools you'll need during your revision would include:

- a planner or diary
- pens
- paper
- highlighters
- flash cards
- and so on ...

Having these tools close to hand will remove simple barriers to successful revision – such as not having a pen!

Revision timetables

Revision timetables are a useful tool to help you organise and structure your work. Remember that the key is to be realistic – don't plan to do too much, or you'll become demoralised.

Revision works best in shorter blocks. So, don't plan to spend two hours solidly revising one topic – you probably won't last that long. Even if you do, it's unlikely the work towards the end of this time will be effective.

If you are making a revision timetable for mock exams (before you've finished your course), you will need to allow time for any homework set in addition to revision.

How to create a revision timetable

Identify the long-term goal and short-term targets you're trying to achieve (and make sure they're SMART). Ask yourself if this a general timetable to use during the term, or one aimed at preparing for a particular exam or assessment. This will affect how you build your plan as your commitments will vary.

Whatever the end goal, don't plan so you only just finish in time. Make sure you plan to cover all the topic areas you need well before the assessment. That way, if you encounter problems that slow you down, you won't run out of time.

Examples of revision timetables

Good example

Revision sessions split into small sections. This helps maintain engagement during the session.

Regular breaks scheduled and realistic expectations of how much revision can be completed in a day.

Times	Mon
8:30am–3:20pm	School
4:00pm–4:30pm	Chemistry (size and mass of atoms)
4:30pm–5:30pm	Football
5:30pm–6:00pm	Dinner
6:00pm–6:30pm	Physics (black body radiation)
6:30pm–7:00pm	Online gaming
7:00pm–7:30pm	Biology (meiosis)

Specific topics given for revision sections – while you don't need to necessarily rigidly stick to this it's good to have a topic focus for each revision session, you can then set targets for the session based around this particular topic area.

Bad example

Unrealistic expectations – timetabling so revision starts at 6:30 am and finishes at 11:00 pm at night is unrealistic and potentially harmful. Failing to achieve set goals can be very demotivating.

Working excessive long hours without adequate sleep and relaxation time can be detrimental to health.

Times	Mon
6:30am–7:20am	Physics
8:30am–3:30pm	School
3:30pm–5:00pm	Physics
5:00pm–7:30pm	Chemistry
7:30pm–11:00pm	Biology

No specific topics mentioned – 'Physics' is far too vague; what areas are they specifically going to work on?

Long blocks of one subject – the student is unlikely to remain engaged for this length of time.

No breaks scheduled – planning breaks, both as a rest and reward, are very important for effective revision.

Tip

Include your other commitments in a revision timetable, such as music lessons, sports, exercise or part-time work. This will give a clearer picture of how much time you have for revision. These commitments could be rewards – they give you something look forward to. Or it may become clear that you may have too much on and have to (temporarily) give something up.

Tip

Make sure you carefully plan how much time you have available before each exam. Miscalculating by even a week could cause problems.

Revision checklist

A revision checklist is an important tool to ensure you are covering all the required specification content. Your teacher may provide you with a revision checklist, but even if they do, making one yourself can be a useful learning activity.

Tip

Some revision guides (like *My Revision Notes*) also have checklists included that you can use.

How to make a revision checklist

1 Read the specification; this is everything you need to know.
2 Split the specification into short statements and place them into a grid.
3 Work through the grid, ticking as you complete each stage for a particular topic. Use practice exam questions to check that your revision has been effective.
4 Return to the areas you are weaker in and focus on improving them.

Example revision checklist

The following is an example statement taken from a GCSE Physics specification. This statement has been used as the basis for an example revision checklist.

Learners should have a knowledge and awareness of the advantages and disadvantages of renewable energy technologies (e.g. hydroelectric, wind power, wave power, tidal power, waste, solar, wood) for generating electricity. Learners should also be able to explain the advantages and disadvantages of non-renewable energy technologies, including fossil fuels and nuclear for generating electricity.

Revision checklist

Specification statement	Covered in class	Revised	Completed example questions	Questions to ask teacher
Advantages and disadvantages of renewable energy resources for generating electricity 1 – hydroelectric, wind power, wave power, tidal power.				
Advantages and disadvantages of renewable energy resources for generating electricity 2 – waste, solar, wood.				
Advantages and disadvantages of fossil fuels for generating electricity.				
Advantages and disadvantages of nuclear power for generating electricity.				

Posters

You could create posters of key processes, diagrams and points and put them up around the house so you can revise throughout the day. Be sure to change the posters regularly so that you don't become too used to them and they lose their impact. See the next section for more on making the most of your learning environment.

Technology

There are many ways to use technology to help you revise. For example, you can make slideshows of key points, watch short videos or listen to podcasts. The advantage of creating a resource yourself is that it forces you to think about a particular topic in detail. This will help you to remember key points and improve your understanding. The finished products should be kept safe so you can revisit them closer to the exam. You could lend your products to friends and borrow ones they've made to share the workload.

Tip

Don't procrastinate by focusing too much on the appearance of your notes. It can be tempting to spend large amounts of time making revision timetables and notes that look nice, but this is a distraction from the real work of revising.

Making your own video and audio

If you record yourself explaining a particular concept or idea, either as a video or podcast, you can listen to it whenever you want. For example, while travelling to or from school. But make sure your explanation is correct, or you may reinforce incorrect information.

Revision slideshows

Slideshows can incorporate diagrams, videos and animations from the internet to aid your understanding of complex processes. They can be converted into video files, printed out as posters, or viewed on screen. It's important to focus on the content of the slideshow – don't spend too long making it look nice.

Social media

Social media contain a wide range of revision resources. However, it is important to make sure resources are correct. If it's user-generated content, there's no guarantee the information will be accurate.

Study vloggers and other students on social media can provide valuable support and a sense of being part of a wider community going through the same pressures as you. However, don't compare yourself to other people in case it makes you feel as if you're not keeping up.

> **Tip**
> ● ● ● ● ● ● ● ● ● ●
> Some students find listening to music helpful when they're revising, even associating certain artists or songs with specific topics. However, music can also be distracting, so only use it if it works for you.

> **Tip**
> ● ● ● ● ● ● ● ● ● ●
> Be aware that social media can also be distracting. It's easy to procrastinate if you're not focused. Advice on reducing distractions can found on page 92.

➤➤ Creating the right environment

The importance of having a suitable environment to revise in cannot be underestimated – you can have the best plan and intentions in the world, but if you're watching TV at the same time, or you can't find the book you need, or you're gasping for a drink and so on, then you're likely to lose concentration sooner rather than later. Make sure you create a sensible working space.

Work area and organisation

It is difficult to concentrate with the distraction of an untidy work area – so keep it tidy! It is also inefficient, as you may spend time looking for things you've mislaid.

The importance of organisation extends to your exercise books and revision folders. You will have at least two years' worth of work to revise and study. Misplacing work can have a negative effect on your revision.

Put together a revision folder with all your notes, practice questions, checklists, timetables and so on. You could organise it by topic so it's easy to find particular information and see the work you have already completed.

Looking after yourself

Revising for exams is a marathon, not a sprint – you don't want to burn out before you reach the exams. Make sure you stay healthy and happy while revising. This is important for your own wellbeing, and helps you revise effectively.

Eat properly

Try to eat a healthy, balanced diet. Keep some healthy snacks nearby so that hunger doesn't distract you when revising. Food high in sugar is not ideal for maintaining concentration, so make sure you're sensible when selecting snacks.

Drink plenty of water

Make sure you have enough water at hand to last your revision session. It's vital to stay hydrated and getting up to get a drink can be a distraction, particularly if you wander past the TV on the way.

Consider when you work most effectively

Different people work better at different times of the day (morning, afternoon, early evening). Try to plan your revision during the times you're most productive. This may take some trial and error at the start of your revision.

Make sure you get enough sleep

Lack of sleep can lead to serious health problems. Late-night cramming is not an effective revision technique.

Avoiding distractions

Social media and other technology can provide an unwelcome temptation when studying. Possible solutions to this distraction include:

Plan specific online activities during study breaks

This could be social media time, videos or gaming. This can also give you something to look forward to while you're working. Be careful to ensure you stick to the allotted break time and don't fall into the trap of 'just one more' video or game.

Switch technology off

Switching the internet off can be the most powerful productivity tool. Turn off your phone and consider avoiding the internet whilst studying, only turning them back on at the end of the study session or during a break. This removes the temptation to constantly check your phone or messages. If you do need access to a device while studying, there are a number of blocker apps and services that can limit what you are able to access.

Tell your family and friends

Make sure people know that you are planning to study for a specific period of time. They'll understand why you may not be replying to their messages and they will help you by staying out of your way. This can also help with positive reinforcement, as you can talk to them afterwards about the successful outcomes of the revision session.

> **Tip**
> Even if you're more productive in the evening, you still need to go to bed early enough to get enough sleep.

» Useful revision techniques

Many students start their GCSE studies with little idea of how to revise effectively. There are many effective revision techniques that are worth trying. And remember, revision is a skill that needs to be learnt and then practised. It may take time to get to grips with some of these strategies, but it will be worth it if you put the effort in.

Memory aids

Before we get onto the revision techniques themselves, here are some tips on how to memorise particularly complicated information. Look out for opportunities to put these techniques into action.

Elaboration

Elaboration is where you ask new questions about what you have already learnt. In doing this you will begin to link ideas together and develop your holistic understanding of the subject. The more connections between topics your brain makes automatically, the easier you will find recalling the relevant information in the exam.

For example, if you have just consolidated your notes on the structure of the plant transport system you might challenge yourself to make a list of all the similarities and differences between the transport systems of plants and humans.

This is useful because, by answering this type of question, your brain will form links between the topics and strengthen your recall while also improving your understanding of both plant and human transport systems.

As part of elaboration you can try and link ideas to real world examples. These will develop your understanding and help you memorise key facts. For example, when revising polymer structure in Chemistry, you could relate this to examples of polymers and how they're used.

Mnemonics

Mnemonics are memory aids that use patterns of words or ideas to help you memorise facts or information. The most common type is where you create a phrase using words whose first letters match the key word or idea or you are trying to learn.

For example, living things can be classified into these taxonomic levels:

- **K**ingdom
- **C**lass
- **F**amily
- **S**pecies.
- **P**hylum
- **O**rder
- **G**enus

A mnemonic to help remember the order of these levels might be:

King **P**hilip **C**ame **O**ver **F**or **G**reat **S**amosas

Other mnemonics include rhymes, short songs and unusual visual layouts of the information you're trying to remember.

Memory palace

Memory palace is a technique that memory specialists often use to remember huge amounts of information. In this technique, you imagine a place (this could be a palace, as in the name of the technique, but it could be your home or somewhere else you're familiar with), and in this location you place certain facts in certain rooms or areas. These facts should, ideally, be associated with wherever you place them, and always stay in the same location and appear in the same order.

You may also find it helpful to 'dress' each fact up in a visual way. For example, you might imagine the information 'gravitational acceleration is ~$10\,m/s^2$' being 'dressed' as the apple that fell on Newton's head. You might then place this apple in the kitchen in your memory palace, 10 metres high on top of one of your cupboards.

Through the process of associating facts and their imaginary location, you are more likely to correctly recall the fact when you come to revisit the 'palace' and locations in your mind.

> **Key term**
>
> Holistic: When all parts of a subject are interconnected and best understood with reference to the subject as a whole.

> **Tip**
>
> It is helpful to use these types of questions to create linked mind maps showing the connections between topic areas.

> **Tip**
>
> When it comes to mnemonics, the sillier the phrase, the better – they tend to stick in your head better than everyday phrases.

Make your revision active

In order to revise effectively, you have to actually *do something* with the information. In other words, the key to effective revision is to make it active. In contrast, simply re-reading your notes is passive and is fairly ineffective in helping people retain knowledge. You need to be actively thinking about the information you are revising. This increases the chance of you remembering it and also allows you to see links between different topic areas. Developing this kind of deep, holistic understanding of the course is key to getting top marks.

Different active techniques work for different people. Try a range of activities and see which one(s) work for you. Try not to stick to one activity when you revise; using a range of activities will help maintain your interest.

> **Key term**
>
> Active revision: Revision where you organise and use the material you are revising. This is in contrast to passive revision, which involves activities such as reading or copying notes where you are not engaging in active thought.

Retrieval practice

Retrieval practice usually involves the following steps.

Step 1 Consolidate your notes

Step 2 Test yourself

Step 3 Check your answers

Step 4 Repeat

Step 1: Consolidate your notes

Consolidating notes means taking information from your notes and presenting it in a different form. This can be as simple as just writing out the key points of a particular topic as bullet points on a separate piece of paper. However, more effective consolidation techniques involve taking this information and turning it into a table or diagram, or perhaps being more creative and turning them into mind maps or flash cards.

Bullet point notes

Here is an example of how you might consolidate bullet point notes from a chunk of existing text.

Original text

Ultrasound waves are inaudible to humans because of their very high frequency. These waves are partially reflected at a boundary between two different media. The time taken for the reflections to echo back to a detector can be used to determine how far away this boundary is, provided we know the speed of the waves in that medium. This allows ultrasound waves to be used for both medical and industrial imaging.

Seismic waves are produced by earthquakes. Seismic P-waves are longitudinal and they travel at different speeds through solids and liquids. Seismic S-waves are transverse, so they cannot travel through a liquid. P-waves and S-waves provide evidence for the structure and size of the Earth's core. The study of seismic waves provides evidence about parts of the Earth far below the surface.

Consolidated notes

- Ultrasound frequency > 20 000, so humans can't hear them
- Ultrasound reflects and the time it takes an echo to return can be used to find distance between target and source
- Ultrasound is used in medicine and industry to obtain images

- Two types of seismic wave in earthquakes: longitudinal P-waves and transverse S-waves
- P-waves can travel through solids and liquids, S-waves through solids only
- Both give information about interior structure of the Earth, e.g. size of the core

Flow diagrams

Flow diagrams are a great way to represent the steps of a process. They help you remember the steps in the right order. An example of a Chemistry flow diagram, for the Haber process, is shown in Figure 4.1.

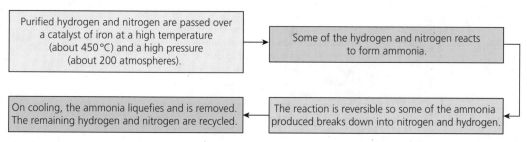

▲ Figure 4.1 **The Haber process**

Mind maps

Mind maps are summaries that show links between topics. Developing these links is a high-order skill – it is key to developing a full and deep understanding of the specification content.

Mind maps sometimes lack detail, so are most useful to make once you have studied the topics in greater detail.

See **Elaboration** (page 93) for more information on the importance of linking ideas in active revision.

> **Key term**
>
> High-order skill: A challenging skill that is difficult to master but has wide ranging benefits across subjects.

Good example of a mind map

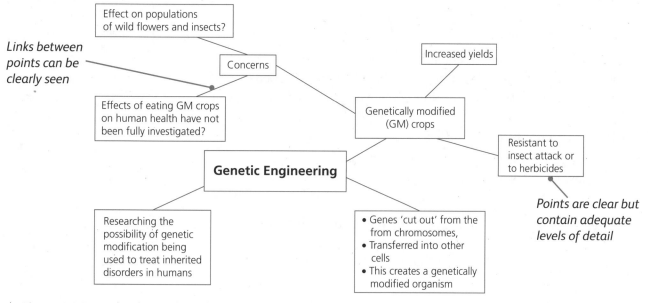

▲ Figure 4.2 **Example of a good mind map**

Bad example of a mind map

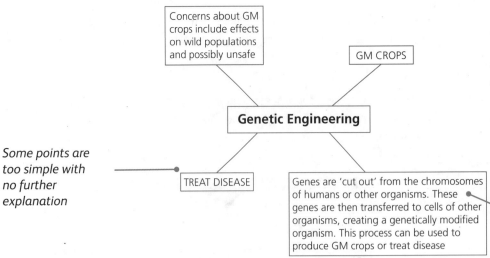

No evidence of linking the points together

Some points are too simple with no further explanation

Points like these are too complex and contain too much text

▲ Figure 4.3 Example of a bad mind map

Flash cards

Flash cards are excellent for things such as key word definitions – write a key word on one side of the card and the definition on the other.

Flash cards can also be used to summarise key points of a process or topic area.

Similar to mind maps, they should be used in conjunction with other revision methods that fully cover the detail required.

Step 2: Test yourself

There is a range of test activities you can do with the notes you've consolidated, including:

● making your own quizzes
● asking friends or family to test you
● picking flash cards randomly from a stack
● trying past exam questions

Whether you've created a test, or are asking other people to test you, it's important that you leave a decent period of time between consolidating your notes and being tested on it. Otherwise, you are not effectively testing your recall.

Step 3: Check your answers

After testing yourself, check your answers using your notes or textbooks. Be hard on yourself when marking answers. An answer that's *almost* right might not gain full credit in an exam. You should always strive to give the best possible answer.

If you get anything wrong, correct your answers on paper (not just in your head). And annotate your answers with anything you've missed along with additional things you could do to improve, such as using more technical language.

Step 4: Repeat

Repeat the whole process, for each topic, at regular intervals. Revisiting activities will help you memorise key aspects and ensure you learn from your previous mistakes. It is especially helpful for topics you find challenging.

Tip
Like with mind maps, do not squeeze too much information on a flash card.

Tip
There are a number of different apps that are useful to help with creating quizzes. Some of these apps also allow you to share these quizzes with friends, so you can help each other out.

Tip
Even though spacing out and mixing up topics are separate sub-sections here, they should be incorporated into your retrieval practice.

When repeating, do not *immediately* revisit the same topic again. Effective revision is more likely if you leave time before revisiting topics you've recently revised, and use this time to mix in other topics.

Spacing out topics

Once you've gone through a whole topic, move on and wait before returning to it and testing your recall. Ideally, you should return to a topic regularly, increasingly long intervals between each return. Returning to a topic needn't take too long – quickly redoing some tests you took before may be enough.

When you return, ask yourself:

- do you know the topic as well as you did when you revised it first time around?
- are you still making the same mistakes?
- what can you improve on?

Identify the key areas you need to go back over.

Allow time for this revisiting process in your revision timetable. Leaving things to the last minute and trying to cram is not an effective way of revising.

> **Tip**
> This is sometimes referred to as 'Spaced Practice'.

Mixing up topics

Mixing up topics (covering a mix of topics during your revision timetable rather than spending long periods of time on one) is an effective revision strategy. It ties into the need to revisit topics at intervals. Mixing up and revising different areas means it's inevitable there will be a space between first revising a topic and then coming back to it a later date.

Studies have shown that, although moving onto different topics more regularly may seem difficult, it could significantly improve your revision. So it's worth persevering.

> **Tip**
> This is sometimes referred to as 'Interleaving'.

» Practise, practise, practise

Completing practice questions, particularly exam-style questions, allows you to apply your knowledge and check that your revision is working. If you're spending lots of time revising but finding you cannot answer the exam questions, then something's wrong with your revision technique and you should try a different one. Examples of practice questions can be found on page 113.

Practice exam questions can be approached in a number of ways.

Complete the questions using notes

This may seem a bit like cheating but it is good, active revision and will show you if there are any areas of your notes that need improving.

Complete questions on a particular topic

After revising a topic area, complete past exam questions on that topic without using your notes. If you find you get questions wrong, go back over your notes before returning to complete questions on this topic area again at a later date. Repeat this process until you are consistently answering all the questions correctly. Annotate your revision notes with points from the mark schemes. More details on the use of mark schemes can be found on page 62.

Complete questions on a topic you have not yet revised fully

This will show you which areas of the topic you know already and which areas you need to work on. You can then revise the topic and go back and complete the question again to check that you have successfully plugged the gaps in your knowledge.

Complete questions under exam conditions

Towards the end of your revision, when you're comfortable with the topics, complete a range of questions under timed, exam conditions. This means in silence, with no distractions and without using any notes or textbooks.

It is important to complete at least some timed activities under exam conditions. The point of this is to prepare you for the exam. Remember, if you spend time looking up answers, talking, looking at your phone and so on, you won't get an accurate idea of timings.

Always ensure you leave enough time to check back over all your answers. Students often lose lots of marks due to silly mistakes, particularly in calculations. These can be avoided by ensuring you check all answers thoroughly.

When working your way up to completing an exam under timed conditions, it can be helpful to begin with timing one or two questions to get yourself used to the speed at which you should be answering them. You can then slowly work your way up to completing full-length papers in the time you would have in the real exam. Make a note of the areas where you found you were spending too long and look for ways to improve.

Effective revision is absolutely vital to success in GCSE Science. As you are studying a linear course you'll be examined on a whole two years' worth of learning. Only by revising effectively and thoroughly can you ensure you have a full and complete understanding of all the content.

Tips

As a guide to timings, you can work out how many marks you should be ideally gaining per minute. To do this, divide the total number of marks available by the time you have in the exam. This will help you get an idea of what questions need longer, but it is not a perfect guide as some questions will take longer than others, particularly the more complex questions that are often found towards the end of the exam paper.

5 Exam skills

You will have spent at least two years learning GCSE physics by the time you take the exam. Given all the work that you have put into your studies, it is important that you know how to apply your knowledge in exam conditions. To get the best grades, you need to understand how to get the maximum mark from every type of question.

This section shows you how to prepare for the exam, how to understand what each question is asking and how to judge the level of content you need to write to get maximum marks.

» General exam advice

Before the exam

Exam specifics

To make sure there are no nasty surprises in the exam, you should read up on how your particular exam board will examine you. If you don't know your exam board or how to access its website or specification, ask your teacher.

Make a note of how many exam papers you will sit, how the papers are split (in terms of content and percentage marks), and the time allowed for each paper. Although unlikely, you should also check whether there are any marks awarded for outside work (for example, some exam boards assess practical work independently). Some of this information is at the front of the specification.

The main section of the specification shows the subject content, which sets out what you must know, understand, and be able to do. Many revision guides, such as *My Revision Notes*, will have checklists for what you need to cover so you can tick them off as you familiarise yourself with each area.

It is helpful to look at past exam papers for your board (sometimes called sample assessment material, or SAMs). Past papers show you the style of questions you can expect – multiple choice, short answer, mathematical, and so on. Check the mark schemes, too, to see how each question is marked. Try using the past papers for practice and revision, and then access the mark scheme to check how well you did.

Planning ahead

Your school will give you your timetable in the summer term, but you can download the full timetable from the exam board's website long before that, if you wish. This will help you plan ahead and make revision schedules.

Closer to the day of the exam, it helps to check that you know when and where your exam is being held. Arrive at least 15 minutes early. Pack your bag the night before to ensure you have everything you need. This will stop you worrying about being under-prepared – which is vital, as getting a good night's sleep before an exam is the best preparation. And don't be fooled into thinking last-minute, late-night cramming will help!

Remember to pack:

- your statement of entry
- pens, pencils and an eraser
- calculator
- ruler (and other mathematical tools you may need)
- see-through pencil case/pouch for the stationery items.

You can't bring your mobile into the exam hall – unless your school has special arrangements, it's best to leave it at home.

During the exam

Understanding what to do

Once you're in the exam room, and have been given the question paper, you should read the advice and instructions on the front cover. Complete the candidate details when told to do so.

As you work through the paper, read each question carefully. Look at command words, which tell you what you have to do.

If it is a long question, plan your answer carefully before you start to write.

If it a mathematical question, think:

1 formula

2 substitutions

3 calculation

4 answer with unit

Look carefully at the space provided for your answer. The amount of space will give you a hint on the maximum you are expected to write to get full marks.

> **Tip**
> Although the amount of lines is a useful guide, do not feel like you have to fill all the space if you are confident that you have written enough for the marks in a shorter space.

Time management

Generally, you should plan to spend no more than one minute per mark. In many cases, you'll need even less than that. If you find yourself spending too long on a question without writing or knowing where to start, mark it with a large asterisk (*) and pass on to the next question. The asterisk will help you find it more quickly when you use any spare time to go back and check unanswered questions.

Showing your working

It is unlikely that you will get everything right and obtain full marks in every single mathematical question. However, in those where you do not get the right answer or score full marks, you can still get partial credit for correct steps along the way. You are much more likely to get these marks if you show the physics you know. Showing your working is essential.

> **Tip**
> You will see in sample mark schemes how examiners give marks for errors carried forward (or ECF). This is a useful indicator of how you can pick up mathematical working marks even when you make a mistake.

Checking your answers

Have a quick look over your answer before moving on to the next question. This helps eradicate any silly mistakes you may have made.

This is particularly true for extended response questions, where up to six marks are available.

Go through a quick checklist in your head. Ask yourself:

- Does each sentence begin with a capital letter and end with a full stop?
- Is the answer set out in paragraphs?
- Is the spelling correct (particularly the technical terms)?
- Are technical terms (weight, mass, energy, power, force, current, and so on) used correctly?

In mathematical questions, check your arithmetic and that your answer is reasonable. If, for example, you are asked 'Calculate the mass of the student…', you should know that an answer of 500 kg (half a tonne) or 5 kg (a bag of potatoes) are both unlikely. If you have an answer like this it is probable that you have incorrectly multiplied or divided by 10 somewhere in the calculation.

Other common troublemakers

Marks in GCSE Physics are often lost needlessly. Here are some common pitfalls:

Pay attention to definitions and laws, which you are expected to recall accurately. Many of the equations you require are provided in the examination. You must be able to recall all the others and be able to use them. A table of equations is provided in the Maths section of this book.

Make sure you can describe all the experiments you carried out during your course, particularly required (or core) practicals. These are often used in six mark extended writing questions.

Make sure you know how to use your calculator. Are you confident, for example, that you can enter numbers in standard index form, find square roots, express numbers to 1 or 2 significant figures and know the meaning of buttons such as $S \Leftrightarrow D$

➤➤ Assessment objectives

Exam board specifications set out the types of questions that can be asked, and the percentage of marks for each question type. Assessment objectives set out how your skills and knowledge will be tested in the exam. There are three of these, and they are spelt out as follows:

Assessment objective	Approximate weighting (%)
AO1 Demonstrate knowledge and understanding of: scientific ideas; scientific techniques and procedures.	40
AO2 Apply knowledge and understanding of: scientific ideas; scientific enquiry, techniques and procedures.	40
AO3 Analyse information and ideas to: interpret and evaluate; make judgments and draw conclusions; develop and improve experimental procedures.	20

AO1 questions

AO1 questions ask you to recall facts, ideas, theories and experiments. These questions are usually worth small amounts of marks (unless you are being asked to recall a lot of separate facts).

(A) Worked example

a State, in order, the stages in the life cycle of a star that is of similar mass to the Sun. [6]

b i Identify what stage the Sun is at in its life cycle. [1]

   ii State what process is occurring now at the core of the Sun. [1]

Model answer

a nebula, protostar, main sequence star, red giant, white dwarf, black dwarf

b i Main sequence   ii Hydrogen fusion to helium.

You will see that these questions are just asking you to state information, without going into the subject matter in any more detail.

AO2 questions

AO2 questions ask you to apply your knowledge to explain scientific ideas, theories, models and to use mathematics. The important word here is 'apply' – this requires you to not only recall information, but use it. AO2 questions tend to reward you with more marks than AO1 questions as they are asking you to do more.

(A) Worked example

Lightning is an example of an electrical discharge. Suggest why during a race a Formula 1 car is earthed using a copper cable before it is refueled.

Model answer

Friction between the rubber tyres and the track causes the car to become charged. Any contact between an earthed conductor (such as a mechanic) and the car will cause an electrical discharge and produce a spark.

This spark could ignite petrol vapour near the car, leading to a fire or an explosion. Earthing the car with a copper cable conducts any possible charge to earth, so sparks will not be produced and refuelling can be done safely.

This is a model answer because it links charging to the friction between the tyres and the road. It identifies the danger – electrical discharge producing a spark and igniting petrol vapour – and it suggests an explanation as to why the remedy stated in the question is likely to be successful.

> **Tip**
>
> Remember that you may be asked to apply your knowledge to new and unexpected situations that may not be clearly indicated in the specification.

AO3 questions

AO3 questions ask you to use your knowledge to develop your own ideas, such as writing hypotheses, designing scientific experiments or suggesting improvements to existing methods. You may also be asked to interpret or evaluate experimental results. AO3 questions are often, but not always, worth the most marks.

Worked example

Describe how you might measure the personal power of a student.

In your method, state the measurements you would make, the measuring instruments you would use, and what you would do with your results to calculate the power. [6]

Model answer

1 First, find the weight of a student, W, using bathroom scales calibrated in newtons.
2 Measure the vertical height, h, of a staircase with a measuring tape.
3 With a stopwatch, measure the time, t, it takes the student to run from the bottom of the staircase to the top.
4 The power, P, developed by the student is given by $P = \dfrac{W \times h}{t}$

This is a model answer because it identifies the three variables to be measured (weight, height of staircase and time), the measuring instruments (bathroom scales, measuring tape and stopwatch) and states what the experimenter has to do.

➤➤ Command words

You should have already come across command words in the Literacy and Revision sections of this book. Command words tell you what you have to do and will often hint at what assessment objective they are testing. For example, 'State' and 'Identify' usually indicate AO1 questions; 'Explain' and 'Calculate' usually indicate AO2 questions; and 'Justify' and 'Outline' usually indicate AO3 questions.

The following is a guide to the most common command words and what they are asking you to do, with examples of what a model answer to each might look like.

> **Tip**
> ....
> At the end of the book, there is a list of command words and their meanings.

Command word: Calculate

Questions that ask you to 'Calculate' want you to use a formula and carry out a calculation (usually with your calculator).

 Worked example

a A nail gun fires a nail of mass 6 g into a piece of wood. The speed of the nail as it enters the wood is 12 m/s. Calculate the kinetic energy of the nail. [3]

b If the average force opposing the nail in the wood is 6 N, calculate how far the nail penetrates the wood. Ignore energy losses. Give your answer to the nearest centimetre. [4]

Model answer

a $K_E = \dfrac{1}{2}mv^2 = \dfrac{1}{2} \times 0.006 \times 12^2 = 0.432\,\text{J}$

b $K_E = F \times d \Rightarrow 0.432 = 6 \times d$

$d = \dfrac{0.432}{6} = 0.072\,\text{m} = 7.2\,\text{cm}$

$d = 7\,\text{cm}$ (to nearest cm)

This is a model answer because it correctly identifies the formulae needed, substitutes in the correct values and completes the calculations accurately. Note how the working has been shown to maximise the chances of picking up marks.

Command word: Choose

Questions that ask you to 'Choose' want you to select information from material that the question supplies.

 Worked example

Some of the main energy resources used to generate electricity are:
- coal
- gas
- geothermal
- nuclear fuel
- oil
- the tides
- wind

Choose from the list those resources that are renewable.

Model answer

The renewable resources are: geothermal, the tides, wind.

This is a model answer because it correctly chooses the information needed from the list. No other information is needed.

Command word: Complete

Questions that ask you to 'Complete' want you to finish something that has already been started in the question. For example, a table or a diagram.

 Worked example

The first part of a nuclear fission reaction is shown below.

Complete the diagram to show how this can start a chain reaction.

You may assume that each fission produces two fission fragments and two fission nuclei.

Nucleus

Neutron → $^{235}_{92}U$

Model answer

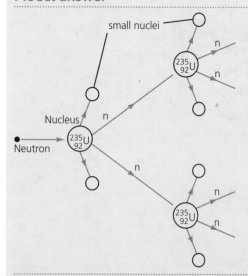

This is a model answer because it completes the diagram that was started, and includes the correct amount of detail and information required.

Command word: Define

Questions that ask you to 'Define' want you state the scientific meaning of a particular word or phrase.

A Worked example

Define what is meant by the specific heat capacity of a material.

Model answer

The specific heat capacity of a substance is the amount of energy required to raise the temperature of one kilogram of the substance by one degree Celsius.

This is a model answer because it provides a simple definition, including correct use and spelling of technical terms.

Command word: Describe

Questions that ask you to 'Describe' want you to give a detailed account, in words, of relevant facts and features relating to the topic being examined. See page 63 for more on how to answer 'Describe' questions.

Command word: Design

Questions that ask you to 'Design' want you to set out a procedure showing how something can be done. See page 67 for more on how to answer 'Design' questions.

Command word: Determine

Questions that ask you to 'Determine' want you to use given data in a question to solve a problem.

A Worked example

Determine how long it would take for the radioactivity of a sample of cobalt-60 to fall from 2560 Bq to 320 Bq if the half-life of the isotope is 5 years.

Model answer

$\frac{2560}{320} = 8 = 2^3$; so, three half-lives are required

$3 \times T_{\frac{1}{2}} = 3 \times 5 = 15 \, \text{years}$

This is a model answer because it correctly determines the data that you are asked to work out. This problem requires mathematical calculation, but a 'Determine' question might just as easily warrant a written answer.

Command word: Draw

Questions that ask you to 'Draw' want you to produce or add to some kind of illustration. This command requires you to take a little more time than that required to produce a 'sketch' (see page 110).

A Worked example

Draw a ray diagram to show how a convex lens can be used to produce an image that is smaller than the object.

Model answer

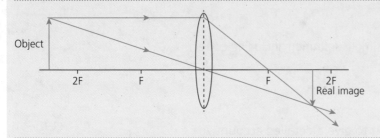

This is a model answer because it is a clear diagram of the image being asked for, with suitable labels and approximate accuracy. If you were asked to 'sketch' an answer instead, you could afford to be a bit more casual, although you should still take as much care in your answer as the time allows.

> **Tip**
> The command word 'Draw' may also be used in questions where you are asked to draw the line of best fit on a graph.

Command word: Estimate

Questions that ask you to 'Estimate' want you to use the numbers given in the question to produce an approximate answer to a problem. See page 19 on how to do an estimate.

A Worked example

The average person in the UK spends 6 minutes 50 seconds in the shower every day. The shower head delivers 6 kg of water per minute at a temperature of 42 °C. The initial temperature of the water is 19 °C.

Estimate how much energy the average person uses to heat the water for a shower, given that the specific heat capacity (SHC) of water is 4200 J/kg °C.

Model answer

Estimated mass of water used, $m = 6\,\text{kg/min} \times 7\,\text{minutes} \cong 40\,\text{kg}$ (1 sf)

Estimated temperature rise, $\Delta\theta = 40°C - 20°C = 20°C$ (1 sf)

SHC of water is 4000 J/kg °C (1 sf)

Estimated energy required, $\Delta E = mc\Delta\theta = 40 \times 4000 \times 20 = 3\,200\,000\,\text{J}$

This is a model answer because it provides an answer in the rough ballpark. You don't need to have an exact answer for 'estimate' questions, although you won't be penalised for using exact figures.

Command word: Evaluate

Questions that ask you to 'Evaluate' want you to use the information supplied as well as your knowledge and understanding to consider evidence for and against an argument. See page 70 for more on how to answer 'Evaluate' questions.

Command word: Explain

Questions that ask you to 'Explain' want you to provide an answer containing some element of reasoning or justification, some of which might be mathematical. See page 65 for more on how to answer 'Explain' questions.

Command word: Give

Questions that ask you to 'Give' want you to provide some new information.

 Worked example

Give a reason why the amount of carbon dioxide in our atmosphere is increasing.

Model answer

We are burning more and more fossil fuels.

This is a simple model answer that answers the question – you do not need to provide any additional information or reasoning unless specified by the question (or if you suspect there are more marks available than you might normally expect for just stating one idea).

Command word: Identify

Questions that ask you to 'Identify' want you to select key information from a source provided for you.

 Worked example

The following represent the nuclei of four different atoms. Note that the A, B, C and D are simply identifiers – they are not the chemical symbols.

$$^{21}_{11}A \quad ^{21}_{12}B \quad ^{22}_{11}C \quad ^{22}_{13}D$$

Identify which two of the nuclei are isotopes of each other.

Model answer

A and C are isotopes.

This is a model answer because, as in 'Choose' questions, you are simply required to pick an answer. Other 'Identify' questions might be more complex than just choosing, for example, questions such as 'Identify the independent variable.' In these questions you may have to recall what an independent variable is and apply it to the information provided. Even so, your answer for any 'Identify' question can be fairly short and to the point.

Command word: Justify

Questions that ask you to 'Justify' want you to provide an argument to support an answer. See page 70 for more on how to answer 'Justify' questions.

Command word: Label

Questions that ask you to 'Label' want you to add text to a diagram, illustration or graph to indicate what particular items are.

A Worked example

The diagram shows a neutral atom. Label the particles identified by the arrows.

Model answer

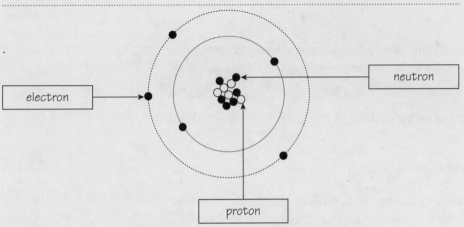

This is a model answer because all the labels are completed clearly. Usually, the labels you have to complete will be signposted as they are in this question, but on rare occasions you may have to draw your own. If this happens, make sure it's really clear what each label is pointing at.

Command word: Measure

Questions that ask you to 'Measure' want you to find a figure of data for a given quantity. On rare occasions you may also be asked to use an instrument to determine a particular property.

A Worked example

The diagram shows a ray of light incident on a plane mirror.

Use the protractor to measure the angle of incidence.

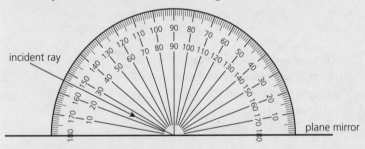

Model answer

Angle of incidence = 64°

This is a model answer because it correctly determines the data that you are asked to work out and includes a unit. In this example, the instrument (protractor) has to be used correctly to calculate the answer, but this is very rare.

Command word: Name

Questions that ask you to 'Name' want you to identify an object, an item, a process, a procedure or a theory.

Worked example

Name the particle responsible for electrical conduction in metals.

Model answer

Electrons

This is a model answer because, as in 'Identify' questions, you are simply asked to select the right information and write it down (either from your own knowledge or from information provided). One or two words is often enough.

Command word: Plan

Questions that ask you to 'Plan' want you to give detailed information about how a procedure or task might be carried out. See page 67 for more on how to answer 'Plan' questions.

Command word: Plot

Questions that ask you to 'Plot' want you to draw and label axes on a grid and mark the points provided. If there is a correlation, you may also be asked to draw the line(s) of best fit. Remember that the line of best fit may be a curve.

 Worked example

A sports car accelerates at 5 m/s^2 uniformly from rest for 5 seconds and then maintains a steady velocity for a further 10 seconds. Plot its velocity–time graph.

Model answer

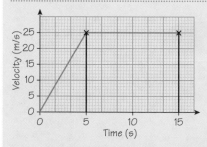

This is a model answer because it identifies the acceleration part of the motion as a line of positive slope, identifies the maximum velocity as 25 m/s and identifies the constant velocity for the last 10 seconds of the motion as a horizontal line.

Command word: Predict

Questions that ask you to 'Predict' want you to write down what you think will happen if a particular condition is met.

 Worked example

As a parachutist falls through the air, the frictional force acting increases as he gets faster and faster. Predict what will happen to his motion when the friction force becomes equal to the weight of the parachutist.

Model answer

By Newton's First Law, he will travel at a steady velocity.

This is a model answer because it provides a plausible outcome using the available information.

Tip

You do not have to provide the reasoning for your prediction unless the question asks for it, but it doesn't hurt to add it in if you have time.

Command word: Show

Questions that ask you to 'Show' want you to demonstrate with clear evidence that the statement given is true. You will often be expected to use the information provided.

(A) Worked example

Below is an energy flow diagram for an electric motor. Show that the efficiency of the motor is 0.8 (80%).

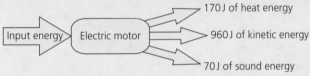

Input energy → Electric motor → 170 J of heat energy
→ 960 J of kinetic energy
→ 70 J of sound energy

Model answer

total energy input = total energy output = (170 J + 960 J + 70 J) = 1200 J

$$\text{efficiency} = \frac{\text{useful energy output}}{\text{total energy input}} = \frac{960\,\text{J [kinetic energy]}}{1200\,\text{J}} = 0.8 = 80\%$$

This is a model answer because it quotes the correct equation, identifies the useful energy output and the total energy input, and correctly carries out the calculation. This would, therefore, be considered suitable evidence for 'Showing' the statement about efficiency to be true.

Command word: Sketch

Questions that ask you to 'Sketch' want you to produce some kind of illustration, which may be produced quickly.

(A) Worked example

Sketch a labelled diagram to show the four forces acting on a crate when it is dragged across a rough wooden floor.

Model answer

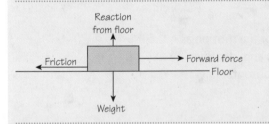

Reaction from floor
Friction
Forward force
Floor
Weight

This is a model answer because it clearly shows all the information requested by the question in a clear fashion. The fact that it is not overly realistic is not penalised because of the 'Sketch' command word.

Command word: Suggest

Questions that begin with the word 'Suggest' require you to use your knowledge and understanding to provide a solution to a problem you are unfamiliar with, or to explain an aspect of physics you may not have studied.

Worked example

Electrical energy is transferred from the power station to the consumer through the National Grid. There has been some success recently in the development of superconductors. These have no electrical resistance.

Suggest and explain how the method of transferring electrical energy to consumers might change if inexpensive materials could be produced, which were superconductors at room temperature. Start by describing the present method of electrical transmission in the grid.

Tip

Note that in this example there are two command words in the question – 'Suggest' and 'Explain'.

Model answer

Currently step-up transformers are used to increase the voltage at the power station.

This means that a given amount of power can be transmitted at a lower current, reducing the amount of energy lost as heat due to the electrical resistance in the transmission lines. However, this voltage is so dangerously high that it must be stepped down by a transformer before distribution to consumers.

If superconductors were used, heat would not be generated in the transmission lines. This might mean that transformers would not be needed in the National Grid.

This is a model answer because it addresses the points in the question in the order required. The suggested change based on the scenario suggested is a plausible one, backed up by scientific knowledge, which is clearly explained.

Command word: Use

Questions that ask you to 'Use' want you to extract appropriate information from diagrams, tables or graphs, etc. Remember that if you do not show how you obtained this information, you will lose marks.

 Worked example

A pebble is thrown into a pond. Graph A shows how displacement changes with time at a fixed distance from where the pebble hits the water. Graph B shows how the displacement, at an instant in time, changes with the distance to the point where the pebble hits the water.

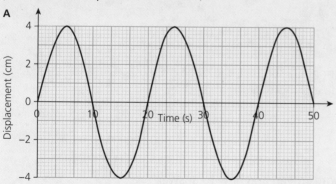

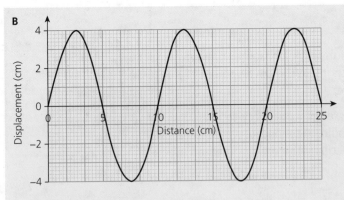

Use the graphs to find:

a the amplitude
b the wavelength
c the frequency
d the speed.

Model answer

a amplitude = maximum displacement from the equilibrium level (horizontal line) = 4 cm

b wavelength = the distance between consecutive peaks (lower graph) = 10 cm

c frequency = number of waves passing per second, and from the upper graph one wave passes every 20 seconds, so $\frac{1}{20}$ th of a wave (or 0.05 waves) pass every second. So, frequency = 0.05 Hz

d speed = frequency × wavelength = 0.05 × 10 = 0.5 cm/s

This is a model answer because each answer clearly uses and interprets the information provided in the question. Note that in some 'Use' questions, outside knowledge may be needed to understand how to use the information given.

Command word: Write

Questions that ask you to 'Write' want you to use written English in your answer. Unlike command words such as 'Design', 'Explain' or 'Describe', 'Write' usually only requires a short response.

A Worked example

Write down how you would calculate the pressure of a body acting on a surface.

Model answer

I would divide the weight of the body by the area of the surface over which this force acts.

This is a model answer because it answers the question in a succinct and clear fashion. For 'Write' questions, you should not include any more information than is required.

Put this into action

Now that you know what all the main command words mean and how to answer them, the next and most important step is to put this learning into action. The next section provides some exam-style practice questions for you to apply your knowledge and help you prepare for the exam. Don't forget that there are also past and sample assessment materials for your specific exam board online.

6   Exam-style questions

>> Paper 1

1 The approximate diameter of an oil molecule can be measured by spreading a thin layer of oil over the surface of a large tray of water.

An oil-drop of volume $0.01\,cm^3$ forms a layer one molecule thick on the surface of a rectangular tray of water measuring 50 cm by 40 cm.

a Calculate the area of the water surface, giving your answer in m^2. [2]

b Calculate the volume of the oil drop in m^3. [1]

c Use your answers to parts **a** and **b** to calculate the diameter of the oil molecule. [2]

d Suggest whether your calculated diameter is likely to be an overestimate or an underestimate of the molecular diameter. [1]

2 a A student is asked to measure the thickness of a sheet of A4 paper. The student measures the thickness of a ream of the paper (500 sheets) and finds it to be 47 mm.

  i Calculate the thickness of one sheet of paper. Give your answer in mm to 2 significant figures. [2]

  ii The student measured the thickness of the ream to the nearest mm.

  Calculate the minimum thickness of one sheet of the paper and explain your reasoning carefully. Give your answer in mm to 2 significant figures. [2]

  b A metal wire is approximately 0.5 mm thick.

  i Describe how the diameter of the wire could be measured accurately using a pencil and a ruler calibrated in mm. [3]

  ii A student is given a long piece of this metal wire.

  In addition to its diameter, suggest other measurements the student needs to make to calculate the density of the metal. [2]

3 Optical fibres are used to carry telephone calls, television programmes and information by broadband. The diagram shows light travelling through an optical fibre.

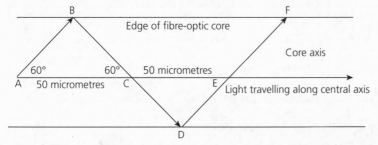

Most of the light is repeatedly reflected and travels along the path ABCDEF.... Some of the light travels along the central axis. The distances AC, CE, and so on, are all $50\,\mu m$ ($50 \times 10^{-6}\,m$). The total length of the core is 900 m.

a Calculate the size of the angles of incidence at B, D and F. [2]

b By considering the triangle ABC, explain why the distance AB = BC = 50×10^{-6} m. [1]

c The speed of light in the core is 1.8×10^8 m/s.

Show that the time taken to travel the distance ABC is twice as long as the time taken to travel the direct distance AC. [2]

d Use your answer to part c to find how much longer the light takes to travel the 900 m length of the core by repeated reflections than by travelling along the central axis. [1]

4 a Of the chemical energy used in a car's engine, $\frac{7}{10}$ is converted into heat. $\frac{9}{10}$ of the remainder is converted into useful energy.

Petrol contains 32 MJ per litre.

A motorist fills her car's petrol tank with 40 litres of fuel.

i Calculate the total energy content of the 40 litres of petrol. [2]

ii Calculate how much of this energy will eventually be converted into heat. [1]

iii Calculate how much energy will be converted into useful energy. [1]

iv Use your answers to parts i and iii to calculate the efficiency of the car's engine. [2]

b The car in part a travels 350 km on a tank containing 20 litres of petrol. A car manufacturer claims that its electric car can go exactly the same distance on a single battery charge of 150 MJ.

i Calculate how much more energy the petrol car uses than the electric car to travel this distance. [2]

ii Explain why you cannot tell from this data alone that the electric car is more efficient. [1]

Governments are encouraging more people to use electric cars, partly because they claim they are better for the environment.

iii Evaluate the advantages and disadvantages of using electric cars rather than petrol cars. [6]

5 a Design an experiment to measure the density of a liquid. [6]

b A student pours different volumes of ethanol into a beaker and then measures the combined mass of the beaker and ethanol. The results are shown in the table.

Combined mass of beaker and liquid M (g)	38	42	46	57	78	82	86	94
Volume of ethanol V (cm³)	10	15	20	50	60	65	70	80

i Plot a graph of M (g) on the y-axis against V (cm³) on the x-axis.

Remember to label each axis, give the appropriate units and apply a suitable scale. [5]

ii With a ruler, draw the line of best fit through the data points. [1]

iii The graph shows an outlier. Define what is meant by an outlier and label it on your graph. [2]

iv Use your graph to find the mass of ethanol in the beaker when it contains 70 cm³ of liquid. [2]

v Calculate the gradient of your graph and give its unit. [3]

6 In a computer simulation, the predicted activity of a particular short-lived radioisotope is tabulated every 30 minutes. The results are shown in this table.

Activity (Bq)	600	476	378	300	238	189	150	119	94	75
Time (hours)	0	0.5	1.0	1.5	2.0	2.5	3.0	3.5	4.0	4.5

a Calculate the probability that a particular nucleus will decay in a given period of 1 hour. [2]

b Plot the graph of *activity* (*y*-axis) against *time* (*x*-axis) and draw a line of best fit. [6]

c From your graph, show that the half-life of this radioisotope is approximately 1.5 hours. [1]

d Estimate the activity of this radioisotope after 1.8 hours. [2]

7 A current of 480 mA enters a network of resistors as shown in the diagram.

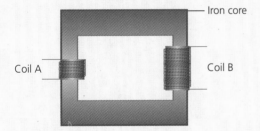

a Complete the table to show the current in each resistor. [4]

Resistance (Ω)	5	7	8	4
Current (mA)				

b Identify which resistor has the greatest voltage across it. Justify your answer. [3]

c Predict what would happen to the current in the 8 Ω resistor if the current in each of the 7 Ω resistors was doubled. Justify your answer. [2]

8 a The diagram shows a transformer, designed to decrease a voltage from 12 kV to 24 V. Coil B has 25 000 turns.

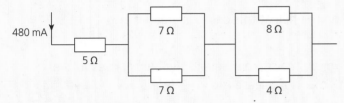

  i Identify which coil, A or B, is the primary coil. Justify your answer. [3]

  ii Calculate the turns ratio of this transformer and use it to find the number of turns in coil A. [3]

b This diagram represents an electricity transmission system, such as the National Grid.

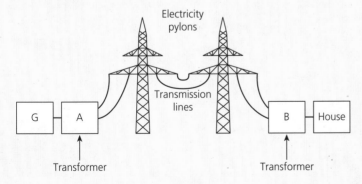

    i State what is represented by box G. [1]

    ii Identify which of the transformers, A or B, is the step-up transformer. [1]

In the Snowdonia National Park in Wales there are no electricity pylons to be seen, but there is an electrical distribution system.

    iii Suggest how the electricity is distributed in Snowdonia, and suggest why this method is not used throughout the country. [3]

9 Nuclear attack submarines are capable of diving to a depth where the water pressure on the hull is $7.35 \times 10^6 \, \text{N/m}^2$.

  a i If the average density of seawater is $1050 \, \text{kg/m}^3$ ($1.05 \, \text{g/cm}^3$), calculate the depth to which such a submarine can submerge. [3]

    ii Suggest a reason why the total pressure on the hull at this depth is greater than $7.35 \times 10^6 \, \text{N/m}^2$. [1]

  b A bubble of volume $0.1 \, \text{cm}^3$ is released in error by the submarine and the bubble rises to the surface. The pressure on the bubble at the moment of release is $6 \times 10^6 \, \text{N/m}^2$.

    i Explain why the volume of the bubble increases as it rises to the surface. [1]

    ii Estimate the volume of the bubble just as it reaches the surface if the pressure on it there is $1 \times 10^5 \, \text{N/m}^2$. [3]

The temperature of the surrounding water at the moment the bubble was released is generally lower than the temperature at the surface.

    iii If the change in temperature was considered, identify whether it would lead to an increase or decrease in the volume estimated in your answer to part **ii**. [1]

[Total = ... / 95 marks]

Answers

» Maths

Arithmetic and numerical computation

Expressions in decimal form (pages 7–8)

Guided questions

1 **Step 1** $v = 1.21\,\text{cm} \times 3.42\,\text{cm} \times 5.63\,\text{cm} = 23.298066\,\text{cm}^3$

   Step 2 volume $= 23.30\,\text{cm}^3$

2 **Step 1** $\dfrac{20\,\text{cm}}{6.4\,\text{s}} = 3.125\,\text{cm/s}$

   Step 2 speed of car $= 3.13\,\text{cm/s}$

Practice questions

3 Max: $210.4\,\text{mm} \times 297.4\,\text{mm}$ Min: $209.5\,\text{mm} \times 296.5\,\text{mm}$

4 area $= 12.2\,\text{mm} \times 15.3\,\text{mm}$
   $= 186.66\,\text{mm}^2 = 186.7\,\text{mm}^2$ (1 dp)

5 Formula for pressure:

   $P = \dfrac{F}{A}$

   $= \dfrac{630}{205}$

   $= 3.07\,\text{N/cm}^2$

   $= 3.1\,\text{N/cm}^2$ (1 dp)

Expressions in standard form (pages 10–11)

Guided questions

1 **Step 1** distance to Moon $= 4 \times 10^8\,\text{m} = 400\,000\,000\,\text{m}$

   time $= 2.592 \times 10^5\,\text{s} = 259\,200\,\text{s}$

   Step 2 speed $= \dfrac{\text{distance}}{\text{time}} = 1543.2\,\text{m/s}$

   Step 3 speed $= 1.5432 \times 10^3\,\text{m/s}$

2 a **Step 1** $200\,\text{m} \times 200\,\text{m} \times 200\,\text{m} = 8\,000\,000\,\text{m}^3$

      Step 2 $8\,000\,000\,\text{m}^3 = 8 \times 10^6\,\text{m}^3$

   b **Step 1** number of atoms $= 8\,000\,000 \times 9\,000\,000$

                       $= 72\,000\,000\,000\,000$ atoms

      Step 2 $= 7.2 \times 10^{13}$ atoms
                       (in standard form)

Practice question

3 distance $=$ speed $\times$ time

         $= 25\,\text{mm per year} \times 500\,000\,\text{years}$

         $= 1.25 \times 10^7\,\text{mm}$

         $= 1.25 \times 10^4\,\text{m}$

Fractions (page 15)

Guided question

1 **Step 1** $60\,\text{kJ} - 10\,\text{kJ} = 50\,\text{kJ}$

   Step 2 $\dfrac{50\,\text{kJ}}{60\,\text{kJ}} = \dfrac{5}{6}$

   Step 3 $\dfrac{3}{4} = \dfrac{9}{12}$

   $\dfrac{5}{6} = \dfrac{10}{12}$

   Step 4 The bigger fraction is $\dfrac{5}{6}$, so the manufacturer's claim is valid.

Practice questions

2 a $1\frac{1}{4} + 3\frac{5}{8} = 4\frac{7}{8}$

   b $2\frac{2}{3} + 4\frac{5}{6} = 7\frac{1}{2}$

   c $7\frac{5}{12} - 6\frac{1}{4} = 1\frac{1}{6}$

   d $3\frac{2}{5} - 4\frac{7}{10} = -1\frac{3}{10}$

   e $2\frac{1}{4} \times 3\frac{5}{8} = 8\frac{5}{32}$

   f $2\frac{2}{5} \times 5\frac{5}{6} = 14$

   g $1\frac{2}{3} \div \frac{4}{9} = 3\frac{3}{4}$

   h $1\frac{1}{2} \div 2\frac{1}{4} = \frac{2}{3}$

3 $\frac{5}{8}$ of the gold is copper

   $\frac{1}{8}$ of the gold $= \frac{1}{5}$ of 95 $= 19$ grams

   All the gold $= \frac{8}{8} = 8 \times 19 = 152$ grams

4 a Sample contains $35\,\text{g}$ salt $+ 965\,\text{g}$ water $= 1000\,\text{g}$ seawater

      Fraction that is salt $= \dfrac{35}{1000} = \dfrac{7}{200}$

   b $1000\,\text{g}$ seawater $= 1\,\text{kg}$ seawater produces 965 pure water $= 0.965\,\text{kg}$ pure water

      $100\,\text{kg}$ seawater produces $100 \times 0.965\,\text{kg} = 96.5\,\text{kg}$ pure water

Ratios (page 17)

Guided question

1 a Step 1 This means that every part in the model is $\frac{1}{20}$th of the size of the corresponding part in the production engine.

b Step 1 length of piston in production engine = 20 times length in model

Step 2 $= 20 \times 5.2\,\text{cm} = 104\,\text{cm}$

c Step 1 length of shaft in model = length of shaft in production engine $\div$ 20

Step 2 $= 1.4\,\text{m} \div 20 = 0.07\,\text{m} = 7\,\text{cm}$

Practice questions

2 mass of beaker : mass of ball = 24 : 18

simplified = 4 : 3

3

Voltage, V (V)	3.2	4.0	4.8	5.6	6.4	7.2
Current, I (A)	0.20	0.25	0.30	0.35	0.40	0.45
Ratio	16 : 1	16 : 1	16 : 1	16 : 1	16 : 1	16 : 1

Voltage is directly proportional to the current since ratio V : I is constant

4 efficiency = useful output energy : total input energy

$$= (3000 - 480) : 3000$$

$$= 2520 : 3000$$

Dividing both numbers by 3000 gives 0.84 : 1 = 0.84

Percentages (pages 18–19)

Guided questions

1 a Step 1 mass of gold = 75% of 1.2 kg $= \frac{75}{100} \times 1.2\,\text{kg}$

$$= 0.9\,\text{kg}$$

b Step 1 percentage of other metals = 100% − 75%
$$= 25\%$$

c Step 1 mass of other metals = 1.2 kg − 0.9 kg = 0.3 kg

2 Step 1 extension = 32 mm − 20 mm = 12 mm

Step 2 fractional extension $= \frac{12\,\text{mm}}{20\,\text{mm}} = 0.6$

Step 3 percentage extension $= 0.6 \times 100\% = 60\%$

Practice questions

3 percentage oil $= \dfrac{80\,\text{g oil}}{(80 + 120)\,\text{g mixture}} \times 100\% = 40\%$

4 a percentage wasted $= \dfrac{\text{wasted energy}}{\text{total input energy}} \times 100\%$

$$= \frac{21\,\text{MJ}}{30\,\text{MJ}} \times 100\% = 70\%$$

b percentage usefully converted =
100% − percentage wasted = 100% − 70% = 30%

5 captured power = 12% of available power

$$= \frac{12}{100} \times 1500\,\text{kW} = 180\,\text{kW} = 180\,\text{kJ/s}$$

energy obtained per minute $= 180\,\text{kJ/s} \times 60\,\text{s} = 10\,800\,\text{kJ}$

Estimating results (pages 19–20)

Guided question

1 Step 1 distance = 100 + 100 + 100 + 100 = 400 m

Step 2 total distance $= 10 \times 400\,\text{m} = 4000\,\text{m}$

Step 3 speed $= \dfrac{\text{distance}}{\text{time}} = \dfrac{4000\,\text{m}}{500\,\text{s}} = 8\,\text{m/s}$

Practice questions

2 estimated fuel use = 20 km per litre

estimated distance per year = 10 000 km

estimated fuel used $= \dfrac{10\,000\,\text{km}}{20\,\text{km/litre}} = 500$ litres

3 estimated distance there and back = 400 000 + 400 000
= 800 000 km = 800 000 000 m

estimated time $= \dfrac{\text{distance}}{\text{speed}} = \dfrac{800\,000\,000\,\text{m}}{3 \times 10^8\,\text{m/s}} = 3\,\text{s}$

(to nearest second)

4 estimated work per second = 200 J/s

estimated time $= (60 \times 60 \times 10)$

$$= 40\,000\,\text{s (1 sf)}$$

estimated work in 10 hours = power × time

$$= 200\,\text{J/s} \times 40\,000\,\text{s}$$

$$= 8\,000\,000\,\text{J}$$

$$= 8\,\text{MJ (to the nearest MJ)}$$

Using sin and sin^{-1} keys (page 21)

Guided question

1 Step 1 $n = \dfrac{1}{\sin c} \Rightarrow c = \sin^{-1}\left(\dfrac{1}{1.52}\right)$

Step 2 $c = 41.1395° = 41.1°$ (1 dp)

Practice questions

2 refractive index $= \dfrac{\sin 90}{\sin 40} = \dfrac{1}{0.6428} = 1.56$ (2 dp)

3 refractive index $= \dfrac{\sin 60}{\sin (90 - 60)} = \dfrac{0.866}{0.500} = 1.73$ (2 dp)

4 refractive index $= \dfrac{\sin i}{\sin r}$, so $2.42 = \dfrac{\sin 70}{\sin r}$

$2.42 \times \sin r = \sin 70$

$\sin r = \dfrac{\sin 70}{2.42} = 0.3883$

$r = \sin^{-1}(0.3883) = 22.8° = 23°$ (to nearest degree)

Handling data

Using significant figures (page 23)

Guided question

1 **Step 1** $V = I \times R$

   Step 2 $V = 1.4 \times 6.8$

   Step 3 $V = 9.52$ volts

   Step 4 Number of sf in data in question is 2.

   Step 5 $V = 9.5$ volts (2 sf)

Practice questions

2 $E = mc\Delta\theta$

   $= 2.55\,kg \times 4200\,J/kg\,°C \times 12.2\,°C$

   $= 130\,662\,J$

   But the number of significant figures in the question is 3, so:

   $\Delta E = 131\,000\,J$ (3 sf)

3 Kinetic energy, $E_k = \dfrac{1}{2}mv^2$

   $= 0.5 \times 0.055 \times 19^2$

   $= 9.9275\,J$

   $= 9.9\,J$ (2 sf)

4 number of molecules in 1 g of water =

   $\dfrac{\text{number of molecules in 18 g}}{\text{18 g mass}}$

   $= \dfrac{6.02 \times 10^{23}}{18} = 3.3444 \times 10^{22}$

   number of molecules = 3.3×10^{22} (2 sf)

Finding arithmetic means (page 24)

Guided question

1 a **Step 1** The outlier is $0.5\,\Omega$.

   b **Step 1** The sum of the other results is:
      $2.1 + 2.2 + 1.9 + 1.8 = 8$

      Step 2 The mean resistance is $8 \div 4 = 2\,\Omega$

Practice questions

2 mean $= \dfrac{(312 + 317 + 313 + 314 + 314)}{5}$

   $= \dfrac{1570}{5} = 314$ mm

3 Mean of 10 values = 4.2.

   So, sum of these 10 values is $4.2 \times 10 = 42$.

   Sum of the given 9 values = $(4.1 + 4.2 + 4.2 + 4.3 + 4.3 + 4.1 + 4.2 + 4.0 + 4.1)$
   $= 37.5$

   So, missing value is $42 - 37.5 = 4.5\,J/g°C$.

Mode and median (page 26)

Guided question

1 **Step 1** 1, 2, 3, 4, 6, 7, 9,

   Step 2 Since the mode is 9, there must be at least two 9s. So, add 9 to the ordered list. There are now 8 numbers in the ordered list.

   Step 3 The median is the fifth number in the ordered list, so the missing number must be 6, 7, 8 or 9. Since the mode is 9, the missing number cannot be 6 or 7. Therefore, the missing number must be 8 or 9.

Practice question

2 a The 20 numbers arranged in order are:

      0.77, 0.77, 0.77, 0.77, 0.77, 0.78, 0.78, 0.78, 0.78, 0.78, 0.78, 0.78, 0.79, 0.79, 0.79, 0.79, 0.79, 0.79, 0.79, 0.79

      The most common number is 0.79 (there are eight of them), so mode = 0.79

   b The numbers in the middle are the tenth and eleventh values. These are both 0.78.

      So, median = 0.78

Frequency tables (page 28)

Practice question

1 a

Number on die	Tally	Frequency
1	ЦНⅠ ||||| 9
2	ЦНⅠ |||| 9
3	ЦНⅠ ЦНⅠ ||	12
4	ЦНⅠ ЦНⅠ ||	12
5	ЦНⅠ |||	8
6	ЦНⅠ ЦНⅠ	10
	Total	60

   b We would expect an unbiased die to show about 10 of each of the numbers 1–6.

      No number comes up more than 12 times or fewer than 8 times. Based on this distribution, there is no strong evidence of bias.

Bar charts (pages 31–32)

Guided question

1 a **Step 1** Female students have a red-coloured bar.

      Step 2 This bar is biggest in 2016.

   b **Step 1** Year in which difference between males and females is greatest is the year in which there is the greatest difference between the height of the bars.

      Step 2 This year is 2016.

c Step 1 Number of male students in 2015 = 80

Step 2 Number of female students in 2015 = 66

Step 3 Total number of students in 2015
= 80 + 66 = 146

Practice question

2 a

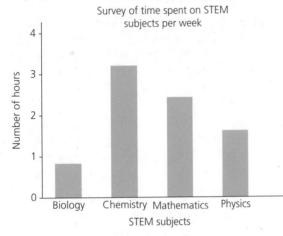

Survey of time spent on STEM subjects per week

b Physics – Biology = 1.6 – 0.8 = 0.8 hours

c total time = (0.8 + 3.3 + 2.4 + 1.6)
= 8.1 hours

Histograms (page 34)

Guided question

1 Step 1 The number of days when the snowfall was between 30 and 40 mm was 6

Step 2 The numbers missing from the middle column of the table are 25, 35, 45 and 55

Step 3 The vertical axis is labelled *Number of days*. The horizontal axis is labelled *Snowfall* (mm) and will range from 0 to 60.

Step 4 The first bar is centre on 15 mm, three days in height and 10 mm wide.

Step 5 Draw the remaining bars. The final bar is two days in height, centred on 55 mm and 10 mm wide.

Step 6 Add the title to the histogram.

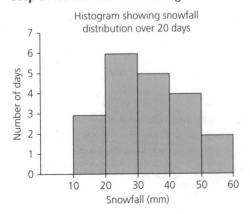

Histogram showing snowfall distribution over 20 days

2 a

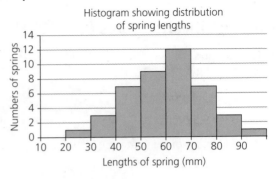

Histogram showing distribution of spring lengths

b Total number of springs =
1 + 3 + 7 + 9 + 12 + 7 + 3 + 1 = 43

c Median class = class containing 22nd spring = greater than 60 but less than 70.

d Modal class = class containing greatest number of springs = greater than 60 but less than 70.

Pie charts (page 36)

Guided question

1 Step 1 The number of students surveyed altogether was: 90

Step 2 Each student in the pie chart is represented by an angle of 4 degrees.

Step 3 So, the angles for each method of transport are:

Walking = 60°; Cycling = 20°; Car = 140°; Bus = 80°; Train = 60°

Step 4 With a compass, draw a large circle to represent the pie.

Step 5 With a ruler, draw a line from the centre of the circle to its circumference.

Step 6 Draw the sectors in the pie using the angles found in Step 3, then label the sectors.

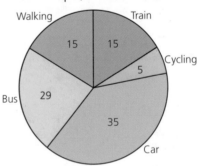

Practice question

2 a Number of people using wood
= 180 – (90 + 45 + 25 + 10 + 2)
= 180 – 172 = 8

b There are 180 people, so each person represents 2°

Energy resource	Gas	Oil	Coal	Electricity	Wood	Other
Number of people	90	45	25	10	8	2
Sector angle (degrees)	180	90	50	20	16	4

c

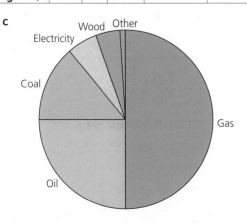

b and c

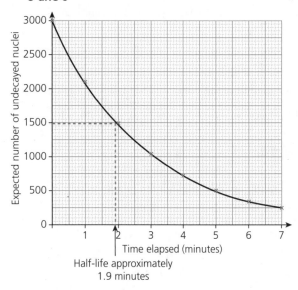

Half-life approximately 1.9 minutes

Simple probability (page 38)

Guided question

1 **a** **Step 1** In one half-life the number of undecayed nuclei falls by 50%.

Step 2 So, in one half-life the 80 million undecayed nuclei will fall to 40 million.

Step 3 From the graph this takes 2 minutes.

b **Step 1** In 6 minutes the number of undecayed nuclei has fallen to 10 million.

Step 2 So, the fraction which have decayed is $\frac{7}{8}$.

Step 3 So, the probability of decay within 6 minutes is 0.875.

Practice question

2 a

Time elapsed (mins)	0	1	2	3	4	5	6	7
Expected number of undecayed nuclei	3000	2100	1470	1029	720	504	353	247

Using a scatter diagram to identify a correlation (pages 40–41)

Guided question

1 **Step 1** No correlation.

Step 2 No trend line can be drawn.

Practice question

2

Distance d of molecule from a fixed point (mm)	0	15	21	26	30	33	38	40
Collision number N	0	1	2	3	4	5	6	7
$\sqrt{N}$	0	**1.0**	1.4	**1.7**	2.0	**2.2**	**2.4**	**2.6**

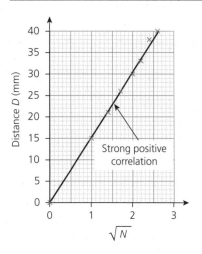

Strong positive correlation

Making order of magnitude calculations (page 42)

Guided question

1 **Step 1** electrical force ÷ gravitational force = 9.22×10^{-8} N ÷ 4.06×10^{-47} N = 2.27×10^{39}

Step 2 The electrical force is 10^{39} times more than the gravitational force.

Practice questions

2 **a** Typical adult is 70 kg so order of magnitude (OM) is 10^2 kg.

b Number of seconds = $24 \times 60 \times 60 = 86400$ s, so OM is 10^5 seconds.

3 For example, 4×10^{-18} m.

4 13.8 billion years = 1.38×10^{10} years, so OM = 10^{10} years.

Algebra

Changing the subject of an equation (page 44)

Guided question

1 **Step 1** $2E = kx^2$

Step 2 $\dfrac{2E}{k} = \dfrac{kx^2}{k}$

Step 3 $\dfrac{2E}{k} = x^2$

Step 4 $\sqrt{\dfrac{2E}{k}} = x$

Step 5 $x = \sqrt{\dfrac{2E}{k}}$

Substituting values into an equation (page 44)

Guided question

1 **Step 1** $a = \dfrac{v - u}{t}$

Step 2 $a = \dfrac{30 - 0}{12}$

Step 3 a = 2.5 m/s^2

Solving simple equations (pages 45–46)

Guided questions

1 **Step 1** $M = Fd$

Step 2 $15 = 6 \times d$

Step 3 d = 2.5 cm

2 **Step 1** $P = rgh$

Step 2 $1\,370\,000 = 1050 \times 10 \times h$

Step 3 $1\,370\,000 = 10\,500 \times h$

Step 4 $h = \dfrac{1\,370\,000}{10\,500} = 130$ metres

Practice questions

3 $v = \lambda$

$3 \times 10^8 = 5 \times 10^{14} \times \lambda$

$\lambda = \dfrac{3 \times 10^8}{5 \times 10^{14}}$

$= 6 \times 10^{-7}$ m

4 $k = \dfrac{F}{x} = \dfrac{7.5\,\text{N}}{2.5\,\text{cm}} = 3$ N/cm

5 average speed $= \dfrac{\text{distance}}{\text{time}} = \dfrac{54\,000\,\text{m}}{(60 \times 60)\,\text{s}} = 15$ m/s

6 $\Delta E = mc\Delta\theta$

$= 2500\,\text{g} \times 4.2\,\text{J/g°C} \times (95 - 25)°\text{C}$

$= 735\,000$ J

$= 735$ kJ

7 $v^2 = u^2 + 2as$ (and, since bag falls from rest, $u = 0$)

$v^2 = 2 \times (-10) \times (-250) = 5000$ where the convention being followed is upwards is positive.

$v = 70.7$ m/s

Inverse proportion (page 47)

Guided question

1 **Step 1** PR = a constant

Step 2 $PR = 1200 \times 48 = 57\,600$

Step 3 $57\,600 = P \times 60$

Step 4 $P = \dfrac{57\,600}{60} = 960$ W

Practice question

2 Since I and d^2 are inversely proportional, the product $I \times d^2$ is a constant.

In this case $I \times d^2 = 1440$.

For the ship: $0.001 \times d^2 = 1440$

a $d^2 = \dfrac{1440}{0.001} = 1\,440\,000$

b $d = \sqrt{1\,440\,000} = 1200$ m

Graphs

Translating between graphical and numerical form (page 49)

Guided question

1 **a** **Step 1** The speed on the vertical axis is half-way between 6 m/s and 8 m/s.

Step 2 At this speed, draw a horizontal line to the graph.

Step 3 From the point where this line meets the graph, draw a vertical line to the time axis.

Step 4 The line meets the time axis at 2.5 seconds. This is the answer.

b **Step 1** The time on the horizontal axis is half-way between 1 s and 2 s.

Step 2 At this speed, draw a vertical line to the graph.

Step 3 From the point where this line meets the graph, draw a horizontal line to the vertical axis.

Step 4 The line meets the speed axis at 9 m/s. This is the answer.

Practice question

2 **a** length of spring when no force is applied = intercept on vertical axis = 20 mm

  b When total length = 70 mm, force = 5 N

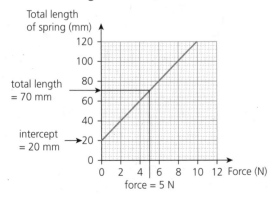

Plotting data on a graph (page 51)

Guided question

1 **Step 1** Draw and label the vertical axis with the letter y and horizontal axis with the letter x.

  Step 2 For the y-axis, the grid is 12 cm high, so each 1 cm distance represents 1 unit.

  For the x-axis, the grid is 12 cm, so each 1 cm distance represents 0.5 units.

  Step 3 The first point is at the intersection where the vertical line at $x = 0$ meets the horizontal line at $y = 4.5$. The second point is at the intersection where the vertical line at $x = 1$ meets the horizontal line at $y = 6$.

  Step 4 Repeat until all points are plotted.

Practice question

2

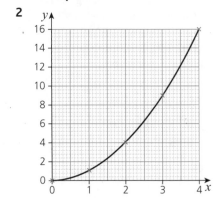

Determining slope and intercept of a straight line (pages 54–55)

Guided question

1 **Step 1** $P = \rho gh + A$

  Step 2 $y = mx + c$

  Step 3 y corresponds to P

    x corresponds to h

    c corresponds to A

    m corresponds to the product ρg

  So, atmospheric pressure = 100 kN/m^2

  Step 4 To find ρ we must divide the gradient by g (or 10 N/kg)

  Step 5 gradient $= \dfrac{\text{rise}}{\text{run}}$

    $= \dfrac{(700 - 100)\text{kN/m}^2}{(50 - 0)\text{m}}$

    $= 12\,\text{kN/m}^3$

    $= 12\,000\,\text{N/m}^3$

  Step 6 $\rho g = 12\,000$ and, since, $g = 10$ N/kg, $\rho \times 10 = 12\,000$

  Step 7 So, the density of the water is

  $\dfrac{12\,000}{10} = 1200\,\text{kg/m}^3$

Practice question

2 **a and b**

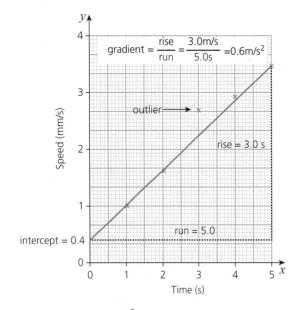

  c gradient = 0.6 m/s^2 and intercept = 0.4 m/s

  d $v = u + at$

  e General equation is $y = c + mx$

    Comparing $y = c + mx$ and $v = u + at$ shows:

    initial speed u corresponds with intercept c, which is 0.4 m/s

    acceleration a corresponds with gradient m, which is 0.6 m/s^2

Geometry and trigonometry

Using and understanding angles (pages 57–58)

Guided question

1

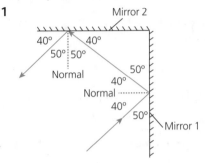

Step 4 angle of reflection at Mirror 2: 50°

Practice question

2

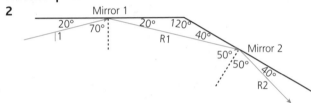

Since angles of incidence and reflection at Mirror 1 are both 70°, the glancing angle at Mirror 1 is 20°.

Since angles in a triangle add up to 180°, the glancing angle at Mirror 2 is 40° and the angle of reflection at Mirror 2 is 50°.

Surface area and volume (page 59)

Guided question

1 **Step 1** volume of cube of size $L = L \times L \times L = L^3$

Step 2 surface area of one face $= L \times L = L^2$

Step 3 total surface area of cube $= 6 \times L^2 = 6L^2$

Step 4 volume : surface area ratio $= \dfrac{L^3}{6L^2} = \dfrac{L}{6}$

which is directly proportional to L.

Practice question

2 Cuboid has

- two surfaces of $1\,\text{cm} \times 2\,\text{cm} = 2 \times (1 \times 2) = 4\,\text{cm}^2$
- two surfaces of $1\,\text{cm} \times 3\,\text{cm} = 2 \times (1 \times 3) = 6\,\text{cm}^2$
- two surfaces of $2\,\text{cm} \times 3\,\text{cm} = 2 \times (2 \times 3) = 12\,\text{cm}^2$

total surface area of cuboid $= (4 + 6 + 12) = 22\,\text{cm}^2$

rate at which each cm^2 radiates energy $= \dfrac{88\,\text{J/s}}{22\,\text{cm}^2}$

$$= 4\,\text{J/s/cm}^2$$

›› Literacy

Extended responses: Describe (pages 63–65)

Expert commentary

1 This is a model answer that would get the full marks.

Connect a variable resistor, an ammeter, a power supply unit (PSU) and a 50 cm length of resistance wire in series with each other. Connect a voltmeter across the resistance wire.

Switch on the PSU and record the readings of current and voltage in a table. Switch off the PSU to allow the wire to cool down. Calculate the resistance by dividing the voltage by the current.

Switch on the PSU again, adjust the variable resistor and record the new current and voltage readings.

Switch off the PSU and calculate the wire's resistance. Then find the average resistance of the two calculations.

Repeat for other wires of increasing length.

Plot a graph of average resistance against length and draw the straight line of best fit. It should pass through (0,0), confirming that the resistance of the wire is directly proportional to its length.

Expert commentary

This description is full of accurate detail and there is clear evidence that the candidate knows exactly what has to be done. A table showing how the results are recorded would have been helpful, but it is not essential.

The student shows knowledge and understanding as to how the apparatus is arranged, the readings that have to be taken and the precautions necessary to obtain satisfactory results. The candidate also knows how to find the resistance and how the results must be processed in order to draw a conclusion.

Peer assessment

2 This question would be awarded a level of 1 and a mark of 2 because only bullets 1, 3 and 6 of the indicative content have been covered.

The student has forgotten to mention volume and they have not mentioned the need to measure the mass of the empty beaker. In steps 1 and 4 they have referred to weight instead of mass. This is an important distinction because density is defined in terms of mass and volume. There is no reference to reliability, and the safety precaution suggested is generic and not specific to this experiment.

Improve the answer

3 This is an improved answer that would get all 6 marks.

White light consists of all the colours of the spectrum: red, orange, yellow, green, blue, indigo and violet. The colour of an object is determined by the colour of the light being shone on it and the colours of the light that the object absorbs.

A red object in white light will appear red, as red is the only colour it will reflect. All the other colours are absorbed.

If a red object is illuminated with light consisting of any mixture of colours that exclude red light, the object will appear to be black.

Extended responses: Explain (pages 65–66)

Expert commentary

1 This is a model answer that would get the full marks.

A transformer consists of a soft iron core on which are wound two coils of wire, called the primary (or input) coil and the secondary (or output) coil. An alternating current applied to the primary coil produces an alternating magnetic field in this coil.

The coils are linked magnetically through the soft iron core, so the alternating magnetic field in the primary coil causes an alternating field at the secondary coil. The alternating field at the secondary coil causes electromagnetic induction in the secondary coil, which results in the appearance of a voltage across the secondary coil.

Because there are more turns of wire in the secondary coil than in the primary coil in a step-up transformer, the voltage across the secondary coil will be greater than that across the primary coil. This is the basis of the turns-ratio

equation: $\dfrac{Np}{Ns} = \dfrac{Vp}{Vs}$

Expert commentary

This explanation is good because it addresses comprehensively all of the items demanded in the question. Reference has been made to construction, type of current, purpose of the core, the turns-ratio equation and electromagnetic induction

Peer assessment

2 This question would be awarded a level of 2 and a mark of 4 because the student gains credit for indicative content bullets 2, 3, 4, and 7.

The student wrongly states that low frequencies are used. There is no indication that the speed of sound in water must be known and the ½ in the equation has been omitted.

Improve the answer

3 This is an improved answer that would get all 6 marks.

Pressure is the force acting on a surface divided by the area of the surface.

The column of liquid is a prism of cross section area, A and height, h.

The volume of liquid is the product $A \times h$.

The mass of the liquid is $A \times h \times \rho$, where ρ is the liquid density.

The weight of the liquid is $A \times h \times \rho \times g$, where g is the gravitational field strength.

The pressure, P, is the weight of liquid divided by the area, so $P = h \times \rho \times g$

Extended responses: Design, Plan or Outline (pages 67–69)

Expert commentary

1 This is a model answer that would get the full marks.

Suspend a spiral spring and a metre ruler vertically using a retort stand, boss and clamp.

Using a ruler, measure the initial length of the spring.

Add a 100 gram (1 N) slotted mass and measure the extended length of the spring.

Repeat for loads up to 6 N in steps of 1 N and record the results in a pre-prepared table.

Calculate the extension for each load, by subtracting the initial length from the new extended length. You can then use this to plot a graph of extension against load. The line of best fit is straight and passes through the origin, indicating direct proportion.

Expert commentary

This is a good answer because it is detailed, yet concise. It is free of irrelevant detail, but contains all the information required to do the experiment. It also demonstrates clearly what has to be done to show direct proportion.

Peer assessment

2 This question would be awarded a level 1 and a mark of 1. This is because credit can be awarded for indicative content bullets 3 and 5 only.

There are also several spelling/grammar mistakes. The candidate has used 'were' instead of 'where', 'mesure' instead of 'measure', 'angel' instead of 'angle' and there are no full stops. The quality of the written English is so poor that the candidate would probably be placed at the bottom of the mark band.

The student does not mention paper, protractor or ray box and therefore gains no credit for indicative points 1, 2 and 4. The angle of refraction is wrongly identified and the marks for points 6 and 7 are therefore lost.

Improve the answer

3 This is an improved answer that would get all 6 marks.

The dependent variable is the upward force on the soft iron plate. The independent variable is the current. The controlled variable is the distance between the electromagnet and the soft iron plate.

You should note the reading (5.0N) when the current is zero. Then, switch on the current and adjust the rheostat until the ammeter reading is 0.5A. Record the results for current and the decreased reading of the newton balance in a table.

Adjust the rheostat to increase the current by 0.5A and record new current and balance readings. Repeat until a set of balance readings and currents, up to 3.0A, have been recorded.

Subtract each balance reading from 5.0N to obtain the upward force from the electromagnet for each current value. You can then plot a graph of upward force from electromagnet against current. If the force produced by the magnet is directly proportional to the current, the graph will be a straight line through the origin.

Extended responses: Evaluate or Justify (pages 70–72)

Expert commentary

1 This is a model answer that would get the full marks.

A positive correlation means that as one variable increases, the other also increases. A negative correlation means that as one variable increases, the other decreases. In this case there is a clear positive correlation because r increases when i increases.

Inverse proportion means that when one variable doubles, the other is halved. This clearly does not occur with this data.

Direct proportion means that the ratio i : r is constant and that when one variable is zero, the other is also zero.

Angle of incidence i (°)	10.0	20.0	30.0	40.0	50.0	60.0	70.0	80.0
Angle of refraction r (°)	6.6	13.2	19.5	25.4	30.7	35.3	38.8	41.0
Ratio i : r	1.5	1.5	1.5	1.6	1.6	1.7	1.8	2.0

The constant ratio (1.5) suggests that for angles of incidence less than 30°, there is direct proportion between i and r, but only up to a limit of proportionality at about i = 30°.

Expert commentary

This is a good answer because it clearly describes the different types of correlation and proportion and then applies these criteria to the data supplied. There is also a recognition that there is limited direct proportion between *i* and *r*.

Peer assessment

2 This question would be awarded a level of 2 and a mark of 4. This is because the student gains credit for indicative content bullets 1, 3, 4 and 5. There are also no spelling or grammar mistakes.

The student refers to what happens at 100 °C and 0 °C when the material changes state. This is irrelevant information because the question specifically refers to liquid water over the range 0 °C to 100 °C.

It is also partly wrong: Ice is less dense than liquid water (that's why icebergs float). The student's attention should have been drawn to the minimum on the graph indicating maximum density at 4 °C.

Improve the answer

3 This is an improved answer that would get all 4 marks.

There is significant absorption of radiation by the thin paper (255 falls to 154) and by the 12 mm lead (149 falls to 74). This implies the presence of two different types of radiation. The absorption by paper indicates the presence of alpha particles in the beam. The lack of significant absorption by the aluminium indicates that there is no beta radiation in the beam. The absorption by the 12 mm lead indicates the presence of gamma radiation.

» Working scientifically

The development of scientific thinking (page 78)

1 A hypothesis is a suggestion made to explain an observation.

2 They carry out experiments.

3 They amend their theory.

4 A student could swallow a battery or be injured by the glass from a broken lamp. [Electric shock is not a suitable answer because batteries up to about 9V have a negligible risk of shocking the user.]

5 The spring could snap or fly around the room, causing serious eye damage. Students should wear goggles or safety glasses.

6 Students could watch a video or do a computer simulation.

7 Burns/scalds from hot water and steam.

8 Mains devices use 230 volts, which is enough to kill. If the cable is damaged or the earthing or fuse is faulty, a student could be electrocuted.

9 a Students might trip on bags in the laboratory.

   b The teacher's action is good practice because it removes the hazard.

10 Peer review involves having research work checked and evaluated by other experts in the same field.

11 This is an issue affecting the whole of society. As such, only the community (through their government) can decide such matters.

Experimental skills and strategies (pages 82–83)

1

	Independent	Dependent	Controlled
a	length	resistance	cross-sectional area (or material from which wire is made)
b	force	acceleration	mass of trolley
c	electrical energy	temperature rise	mass of aluminium
d	vertical height	time	mass of marble
e	mass	periodic time	spring constant
f	current	heat produced per second	wire's resistance

2 a Hypothesis: The greater the load W, the further the 5 N load must be placed from the midpoint of the rod to restore equilibrium.

b When the rod comes to rest in a horizontal position.

c No. The hypothesis is valid but it does need to be modified. For example, you might add to the end: '... provided the load W is less than 10 N.'

3 a top-pan balance

b burette

c stopwatch (digital or analogue)

d milliammeter (digital or analogue)

e liquid-in-glass thermometer

4 a range = $10.1 - 9.7 = 0.4 \, \text{m/s}^2$

mean = $(9.7 + 9.8 + 9.8 + 9.8 + 10.1) \div 5 = 9.8 \, \text{m/s}^2$

b Taking the mean reduces the effect of results that are too small or too big.

5 Repeating measurements increases reliability.

6 a A suitable format might be:

Volume (cm^3)	20	35	45	50	55
Mass (g)	16	28	36	40	42

b The student might want to repeat the 55 cm^3 and 42 g results, because all other pairs of values give a density of 0.80 g/cm^3, but this pair of values gives a density of 0.76 g/cm^3.

7 a Stopclock

b i Student C

ii Student A

c It will improve reliability.

Analysis and evaluation (page 86)

1 a Student A: Mean is 5.6 Ω, and the uncertainty is 0.3 Ω

Student B: Mean is 5.7 Ω (1 dp), and the uncertainty is 0.1 Ω (1 dp)

b A is more accurate than B because the mean value is closer to the true value.

2 a Systematic

b Random

c Neither – this is a mistake.

3 An error is caused by a defect in the apparatus, a problem with the experimental technique or inconsistency in the measurement. A mistake is caused by a person using apparatus (for example, a calculator) incorrectly.

4 a systematic

b systematic

5 It reduces random error where some results are too big and others too small. Repeating and averaging cancels out the small values with the big values.

6 a All the results are smaller than the true value, so this is a systematic error.

b Reproducible and valid

⟫ Exam-style questions

Paper 1 (pages 113–116)

1 a area = length × breadth

$= 0.5 \, \text{m} \times 0.4 \, \text{m}$ [1]

$= 0.2 \, \text{m}^2$ [1]

b volume = $0.01 \, \text{cm}^3 = 0.01 \div 1\,000\,000$

$= 1 \times 10^{-8} \, \text{m}^3$ [1]

c diameter = volume ÷ area

$= 1 \times 10^{-8} \div 0.2$ [1] (allow error carried forward)

$= 5 \times 10^{-8} \, \text{m}$ [1]

d The oil on the surface of the water cannot be *less* than 1 molecule thick, suggesting that the measured diameter is an overestimate. [1]

2 a i thickness = 47 mm ÷ 500 [1] = 0.094 mm [1]

ii By measuring to the nearest mm, the thickness of the 500 sheet ream t is known to be $46.5 \leqslant t < 47.5$ mm. [1]

The minimum thickness of a single sheet = (46.5 mm) ÷ 500 = 0.093 mm (to 2 dp). [1]

b i Wrap about 20 turns of wire on a pencil. Push the turns of wire together to form a tight coil [1]. Measure the length of the coil with a ruler [1]. Divide the length by the number of turns to find the wire's thickness [1].

ii Mass of wire [1] and length of wire [1].

3 a Award one mark for one or two correct angles correctly calculated, and two marks for all three angles correctly calculated:

Angle sum in the triangle ABC is 180°, so angle at B must be 60°. Angle of incidence = half this angle = 30°. Angles of incidence at B and C are also 30°. [2]

b ABC is an equilateral triangle, so distance $AB = BC = 50 \times 10^{-6}$ m [1]

c Distance A to B to C is $(50 + 50) = 100\,\mu$m [1], which is twice the direct distance AC. Therefore, the time taken to travel the distance ABC will be twice as long as the time taken to travel the direct distance AC [1].

d Time to travel 900 m = distance ÷ speed

$$= 900\,\text{m} \div 1.8 \times 10^{8}\,\text{m/s [1]}$$

$$= 5\ \text{microseconds [1]}$$

Additional time by repeated reflection = 5 microseconds [1]

4 a i energy content = 40 litres × 32 MJ per litre [1]
= 1280 MJ [1]

ii energy converted to heat = $(7 \div 10) \times 1280$ MJ
= 896 MJ [1] (allow error carried forward).

iii useful energy = $0.9 \times (1280 - 896)$
= 345.6 MJ [1] (allow error carried forward).

iv efficiency = useful output energy ÷ total input energy = $345.6 \div 1280$ [1] = 0.27 [1] (allow error carried forward).

b i 20 litres petrol contains 0.5×1280 MJ = 640 MJ

additional energy in petrol = $(640 - 150)$ MJ [1]
= 490 MJ [1] (allow error carried forward).

ii We do not know the mass of the car or its load carrying capacity. [1]

iii Award one mark for each of the indicative content points covered below, up to a maximum of six marks. Accept any other reasonable answers.

Disadvantages

- Batteries have a much lower energy density than petrol (they store fewer joules per kilogram), so battery cars have a much smaller range than petrol cars.

- Batteries for electric cars are still being developed, so electric cars are more expensive than petrol cars as manufacturers try to recover the development costs.

- It is said that battery cars are less polluting than petrol cars, but this takes no account of the additional pollution that might be created by producing the electricity to drive them. (It is thought that if all the cars in the UK were electric cars, the UK would require the equivalent of 10 additional nuclear power stations for these cars alone.)

- Battery cars use rare earth metals from the Earth's crust and there is currently no known way to recycle them.

Advantages

- Electric cars produce no CO_2, so no greenhouse gases are produced by the cars themselves.

- No oxides of nitrogen or particulate pollutants are produced by electric cars – these pollutants cause serious health problems.

- The engine of an electric car is much more efficient than that of a petrol car, so less of the Earth's resources are consumed by the cars themselves.

5 Using a top-pan balance, find the mass M_1 of a measuring cylinder and record it in a table [1]. Pour some liquid into the cylinder and record its volume V [1]. Measure the combined mass, M_2, of the measuring cylinder and liquid [1]. Calculate the value of $(M_2 - M_1) \div V$ [1]. This is the liquid's density [1]. For reliability, repeat the experiment five more times and find the average density [1].

b i See graph. Axes drawn and labelled correctly [1]; Suitable scales chosen [1]; Units used correctly [1]; All points plotted correctly [2].

ii See graph. Award one mark for line of best fit drawn correctly.

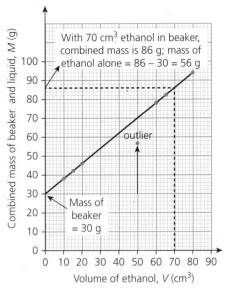

iii An outlier is a value that falls well outside the range of the other values [1] and is not included as valid because it would lead to an incorrect conclusion. Outlier is (50, 57) – see graph [1].

iv Correctly use graph to identify values [1]

mass of ethanol =
86 g – 30 g (combined mass – mass of beaker)
= 56 g [1]

v gradient = rise ÷ run [1]
$= (94 - 30)\,\text{g} \div (80 - 0)\,\text{cm}^3$ [1]
$= 0.8\,\text{g/cm}^3$ [1]

6 a $P = (600 - 378) \div 600$ [1] $= 0.37$ [1]

b Both axes labelled with units as in the table [1]; Scales chosen to cover at least half of each axis [1]; Scales chosen allow for a straightforward interpolation [1]; Points plotted to within 1 small square [2] (deduct ½ mark for each incorrect or missing point – round up mark if necessary); Smooth curve drawn through the data points [1].

c See graph for working.

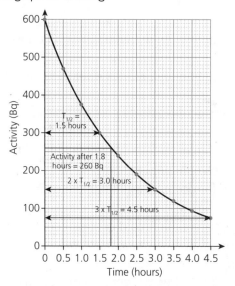

d Evidence from graph – vertical line at 1.8 hours to curve, horizontal line to vertical axis. [1]

Activity = 260 Bq [1]

7 a Award one mark for each correctly calculated current.

Resistance / Ω	5	7	8	4
Current / mA	480	240	160	320

b There are three parts to the network:

1 a single resistor of 5 Ω

2 two 7 Ω resistors, with combined resistance of 3.5 Ω [1]

3 an 8 Ω resistor and a 4 Ω resistor, with combined resistance of 2.67 Ω [1]

The voltage is greatest across the combination which has greatest resistance – the single 5 Ω resistor [1].

c Total current in network doubles, so current in 8 Ω resistor also doubles. [2]

8 a i B is the primary coil [1]. The transformer reduces the voltage from 12 kV to 24 V and is, therefore, a step down transformer [1]. Step-down transformers have more turns on the primary coil than the secondary coil [1].

ii turns ratio = voltage ratio = $V_s : V_p = 24 : 12\,000$
$$= 1 : 500 \text{ [1]}$$
$$N_s = N_p \times \frac{V_s}{V_p}$$
$$= 25\,000 \times \frac{1}{500} \text{ [1]}$$
$$= 50 \text{ turns [1]}$$

b i Generator (in power station) [1]

ii A [1]

iii In Snowdonia, the electricity is transmitted underground [1] because the pylons are considered to be unsightly and would spoil the magnificent scenery [1]. This is not usually done because underground transmission is so expensive [1].

9 a i $P = \rho \times g \times h$ [1]
$$7.35 \times 10^6 = 1050 \times 10 \times h \text{ [1]}$$
$$h = 700 \text{ m [1]}$$

ii The total pressure also includes the pressure of the air above the water. [1]

b i Volume increases because the pressure decreases. [1]

ii By Boyle's Law: $p_1 V_1 = p_2 V_2$ [1]

So $(6 \times 10^6 \times 0.1) = (1 \times 10^5 \times V_2)$ [1]

which gives:
$$V_2 = \frac{6 \times 10^5}{1 \times 10^5}$$
$$= 6 \text{ cm}^3 \text{ [1]}$$

iii An increase in water temperature would lead to an increase in the bubble volume in accordance with the gas laws. [1]

Key terms

Accuracy: Accuracy is how close we get to the true value of any physical measurement.

Active revision: Revision where you organise and use the material you are revising. This is in contrast to passive revision, which involves activities such as reading or copying notes where you are not engaging in active thought.

Angle of incidence, i: Angle between the incident ray and the normal to the boundary of a transparent material.

Angle of reflection: Angle between a reflected ray and the normal.

Angle of refraction, r: Angle between the refracted ray and the normal to the boundary of a transparent material.

Arithmetic mean: The sum of a set of values divided by the number of values in the set – it is sometimes called the average.

Bar charts: Charts showing discrete data in which the height of the unconnected bars represents the frequency.

Base units: The units on which the SI system is based.

Categoric variables: Variables that are not numeric (such as colour, shape).

Causal relationship: The reason why one quantity is increasing (or decreasing) is that the other quantity is also increasing (or decreasing).

Continuous data: Data that can take any value within a range, such as the mass of a beaker.

Continuous variables: The variables that can have any numerical value (such as mass, length).

Controlled variables: The variables that are kept constant throughout an experiment.

Critical angle, c: The angle of incidence in an optically dense medium when the angle of refraction in air is 90°.

Decider figure: The integer after the number of decimal places required that decides whether we must round up or not.

Decimal places: The number of integers given after a decimal point.

Denominator: The number below the line in any fraction.

Dependent variable: The variable that changes because of the change in the independent variable.

Derived units: Combinations of base units such as m/s and kg/m^3.

Direct proportion: Quantities are directly proportional if they are in a constant ratio to each other.

Discrete data: Data that can only have particular values, such as the number of marbles in a jar.

Ethical issues: Issues that involve deciding whether a course of action is morally right or wrong.

Fair test: A test in which there is one independent variable, one dependent variable and all other variables are controlled.

Grouped frequency table: A table of data in which items are ranged in classes or groups.

High order skill: A challenging skill that is difficult to master but has wide ranging benefits across subjects.

Histograms: Charts showing continuous data in which the area of the bar represents the frequency.

Holistic: When all parts of a subject are interconnected and best understood with reference to the subject as a whole.

Hypothesis: A suggestion made, with limited evidence, to explain a phenomenon or observation that can then be tested through practical means. The plural of hypothesis is hypotheses.

Incident ray: A ray that strikes a surface.

Independent variable: The variable that the physicist decides to change.

Index: Index is the power to which a number or letter is raised. The plural of index is indices.

Integers: These are whole numbers, which includes zeros.

Inversely proportional: Quantities x and y are inversely proportional to each other if their product xy is constant.

Leading zeros: Zeros before the first significant figure in small numbers; for example, there are two leading zeros (the zeros before the 2) in 0.002034.

Multiples: Large numbers of base or derived units, such as kilo- in kilogram.

Negative correlation: This occurs if one quantity tends to decrease when the other quantity increases.

No correlation: There is no relationship whatever between two quantities.

Normal: A line drawn at right angles to a surface.

Numerator: The number above the line in any fraction.

Order of magnitude: If we write a number in standard form, the nearest power of 10 is its order of magnitude.

Outlier: A value that 'lies outside' the other values in a set of data observations, either because it is much higher or much lower.

Peer review: The process in which experts in the same area of study read, consider and report on the findings of another scientist before it is considered for inclusion in a scientific journal (magazine read by scientists).

Percentage (%): A fraction of 100.

Phenomenon: An observation that prompts you to ask questions. The plural of phenomenon is phenomena.

Positive correlation: This occurs if one quantity tends to increase when the other quantity increases.

Precision: Precision measures the extent to which measurements are the same.

Random error: An error that causes a measurement to differ from the true value by different amounts each time.

Ratio: A way to compare quantities; for example, three apples and four oranges are in the ratio 3:4.

Recurring: When a number goes on forever.

Reflected ray: A ray that is reflected from a surface.

Refractive index: The ratio $\sin i : \sin r$.

Reliable: Results are reliable if they are valid, repeatable and reproducible.

Repeatable: Results are repeatable if similar results are obtained when an experiment is repeated by the same person several times.

Reproducible: Results are reproducible if similar results are obtained when the experiment is repeated by another person or by using a different technique.

Resolution: Resolution is the fineness to which an instrument can be read.

Scatter graph: A graph plotted between two quantities to see if there might be a relationship between them.

Significant figures (sf): Approximations to a number, determined by a set of mathematical rules.

Standard form: A number in the form $a \times 10^n$ used when writing down very large or very small numbers.

Submultiples: Fractions of a base unit or derived unit, such as centi- in centimetre.

Systematic error: An error that causes a measurement to differ from the true value by the same amount each time.

Trailing zeros: Zeros after the first significant figure; for example, the zero to the right of the 2 in 0.002 034 is a trailing zero and would be significant if expressing this number to 3 sf.

Ungrouped frequency table: A table of data in which each item is discrete.

Valid: Results are valid when the measurements are correct measures of the property being investigated. For example, measuring the length of a magnet is not a valid way to measure its strength.

Command words

Calculate: Questions that ask you to 'Calculate' want you to use a formula and carry out a calculation (usually with your calculator).

Choose: Questions that ask you to 'Choose' want you to select information from material that the question supplies.

Complete: Questions that ask you to 'Complete' want you to finish something that has already been started in the question. For example, a table or a diagram.

Define: Questions that ask you to 'Define' want you state the scientific meaning of a particular word or phrase.

Describe: Questions that ask you to 'Describe' want you to give a detailed account, in words, of relevant facts and features relating to the topic being examined.

Design: The command word 'Design' wants you to use your knowledge and experience and be creative in solving an experimental task.

Determine: Questions that ask you to 'Determine' want you to use given data in a question to solve a problem.

Draw: Questions that ask you to 'Draw' want you to produce or add to some kind of illustration. This command requires you to take a little more time than that required to produce a 'sketch'.

Estimate: Questions that ask you to 'Estimate' want you to use the numbers given in the question to produce an approximate answer to a problem.

Evaluate: The command word 'Evaluate' requires you to use the information supplied in the question, as well as any relevant outside knowledge, to consider evidence for and against an argument.

Explain: The command word 'Explain' means that the answer must contain some element of reasoning or justification.

Give: Questions that ask you to 'Give' want you to provide some new information.

Identify: Questions that ask you to 'Identify' want you to select key information from a source provided for you.

Justify: The command word 'Justify' means you need to use the evidence supplied to support and take one argument forward.

Label: Questions that ask you to 'Label' want you to add text to a diagram, illustration or graph to indicate what particular items are.

Measure: Questions that ask you to 'Measure' want you to find a figure of data for a given quantity. On rare occasions you may also be asked to use an instrument to determine a particular property.

Name: Questions that ask you to 'Name' want you to identify an object, an item, a process, a procedure or a theory.

Outline: The command word 'Outline' means 'summarise', but it is often used to ask students to set out how something should be done (such as 'Outline a plan', 'Outline an experiment', and so on).

Plan: Questions that ask you to 'Plan' want you to give detailed information about how a procedure or task might be carried out.

Plot: Questions that ask you to 'Plot' want you to draw and label axes on a grid and mark the points provided. If there is a correlation, you may also be asked to draw the line(s) of best fit. Remember that the line of best fit may be a curve.

Predict: Questions that ask you to 'Predict' want you to write down what you think will happen if a particular condition is met.

Show: Questions that ask you to 'Show' want you to demonstrate with clear evidence that the statement given is true. You will often be expected to use the information provided.

Sketch: Questions that ask you to 'Sketch' want you to produce some kind of illustration, which may be produced quickly.

Suggest: Questions that begin with the word 'Suggest' require you to use your knowledge and understanding to provide a solution to a problem you are unfamiliar with, or to explain an aspect of physics you may not have studied.

Use: Questions that ask you to 'Use' want you to extract appropriate information from diagrams, tables or graphs etc. Remember that if you do not show how you obtained this information, you will lose marks.

Write: Questions that ask you to 'Write' want you to use written English in your answer. Unlike command words such as 'Design', 'Explain' or 'Describe', 'Write' usually only requires a short response.

Formulae

This table lists the equations in the specifications of the main GCSE exam boards. Most of them you have to remember and be able to use, but use the guide below to check for your board's specifics.

KEY
RR = required recall (you must remember the equation **and** be able to use it).
S&U = select and use (you will be given a list of formulae and you need to be able to select the correct one and use it – but you do not need to remember it).
N/A = not required for this specification.
DS = formula given on data sheet in examination.
Highlight = the equation will only appear in Higher Tier papers.

Context	Equation	AQA	OCR	Edexcel UK	WJEC	CCEA	Edexcel International
Weight	$W = mg$	RR	RR	RR	RR	RR	RR
Work	$W = Fs$ or $E = Fd$	RR	RR	RR	RR	RR	RR
Hooke' Law	$F = ke$ or $F = kx$	RR	RR	RR	RR	RR	RR
Moment	$M = Fd$	RR	RR	RR	RR	RR	S&U
Pressure	$p = \frac{F}{A}$	RR	RR	RR	S&U	RR	RR
Distance	$s = vt$	RR	RR	RR	RR	RR	RR
Average speed	$\bar{v} = \frac{1}{2}(u + v)$	RR	N/A	N/A	RR	RR	RR
Acceleration	$a = \frac{\Delta v}{t}$	RR	RR	RR	RR	RR	RR
Newton's Law	$F = ma$	RR	RR	RR	RR	RR	RR
Momentum	$p = mv$	RR	RR	RR	RR	N/A	S&U
Kinetic Energy	$E_k = \frac{1}{2}mv^2$	RR	DS	RR	RR	RR	RR
GPE	$E_p = mgh$	RR	RR	RR	RR	RR	RR
Power	$P = \frac{E}{t}$ or $P = \frac{W}{t}$	RR	RR	RR	RR	RR	RR
Efficiency	efficiency = $\frac{\text{useful output energy}}{\text{total input energy}}$	RR	RR	RR	S&U	RR	RR
	efficiency = $\frac{\text{useful output power}}{\text{total input power}}$	RR	N/A	N/A	S&U	RR	N/A
Wave equation	$v = f\lambda$	RR	RR	RR	S&U	RR	RR
Charge	$Q = It$	RR	RR	RR	N/A	RR	RR
Ohm's Law	$V = IR$	RR	RR	RR	S&U	RR	RR
Series resistance	$R_T = R_1 + R_2$	RR	RR	RR	S&U	RR	N/A
Parallel resistance	$\frac{1}{R_T} = \frac{1}{R_1} + \frac{1}{R_2}$	N/A	N/A	N/A	S&U	RR	RR
Joule's Law (power)	$P = VI$	RR	RR	RR	RR	RR	RR
Joule's Law (power)	$P = I^2R$	RR	RR	RR	RR	N/A	RR
Power	$E = Pt$ or $E = IVt$	RR	RR	DS	S&U	RR	RR
Voltage	$E = QV$	RR	RR	RR	N/A	RR	RR
Density	$\rho = \frac{m}{v}$	RR	RR	RR	S&U	RR	RR
Pressure	$P = h\rho g$	RR	DS	DS	RR	N/A	RR
Motion equation	$v^2 = u^2 + 2as$	RR	DS	DS	RR	N/A	RR
Force (due to an impulse)	$F = \frac{m\,\Delta v}{\Delta t}$	RR	N/A	DS	RR	N/A	S&U
Heat capacity	ΔE (or Q) $= mc\Delta\theta$	RR	DS	DS	S&U	N/A	S&U
Periodic time	$T = \frac{1}{f}$	RR	N/A	N/A	N/A	RR	RR

Context	Equation	AQA	OCR	Edexcel UK	WJEC	CCEA	Edexcel International
Magnification	$M = \dfrac{h_{image}}{h_{object}}$	RR	N/A	N/A	N/A	N/A	N/A
Force in magnetic field	$F = BIl$	RR	DS	DS	RR	N/A	N/A
Latent Heat	$E = mL$ or $Q = mL$	RR	DS	DS	RR	N/A	N/A
Transformer turns-ratio	$\dfrac{V_p}{V_s} = \dfrac{n_p}{n_s}$	RR	DS	DS	RR	RR	S&U
Transformer power	$V_s I_s = V_p I_p$	RR	DS	DS	N/A	N/A	S&U
Boyle's Law	$pV = $ constant or $p_1 V_1 = p_2 V_2$	RR	DS	DS	RR	N/A	RR
Pressure Law	$\dfrac{p_1}{T_1} = \dfrac{p_2}{T_2}$	N/A	N/A	N/A	N/A	N/A	RR
Gas Law	$\dfrac{PV}{T} = $ a constant	N/A	N/A	N/A	S&U	N/A	N/A
Energy stored in spring	$E = \dfrac{1}{2}kx^2$	DS	DS	DS	RR	N/A	RR
Snell's Law	$.n = \dfrac{\sin i}{\sin r} = \dfrac{1}{\sin c}$	N/A	N/A	N/A	N/A	N/A	RR